# Collins

# Maths Progress Tests for White Rose

Year 3/P4

# Sarah-Anne Fernandes

William Collins' dream of knowledge for all began with the publication of his first book in 1819. A self-educated mill worker, he not only enriched millions of lives, but also founded a flourishing publishing house. Today, staying true to this spirit, Collins books are packed with inspiration,
innovation and practical expertise. They place you at the centre of a world of possibility and give you exactly what you need to explore it.

Collins. Freedom to teach.

Collins
An imprint of HarperCollins*Publishers*
The News Building
1 London Bridge Street
London
SE1 9GF

1st Floor
Watermarque Building
Ringsend Road
Dublin 4
Ireland

Browse the complete Collins catalogue at
www.collins.co.uk

© HarperCollinsPublishers Limited 2019

10 9 8 7 6 5 4 3 2 1

ISBN 978-0-00-833353-9

All rights reserved. No part of this publication may be reproduced, stored in a retrieval system,
or transmitted in any form by any means, electronic, mechanical, photocopying, recording or otherwise, without the prior written permission of the Publisher or a licence permitting restricted copying in the United Kingdom issued by the Copyright Licensing Agency Ltd., 5th Floor, Shackleton House, 4 Battle Bridge Lane, London, SE1 2HX.

British Library Cataloguing in Publication Data. A catalogue record for this publication is available from the British Library.

Author: Sarah-Anne Fernandes
Publisher: Katie Sergeant
Commissioning Editor: Fiona Lazenby
Product Developer: Mike Appleton
Copyeditor: Joan Miller
Proofreader: Catherine Dakin
Design and Typesetting: Ken Vail Graphic Design
Cover Design: The Big Mountain Design
Production controller: Katharine Willard

# Contents

**How to use this book**      iv

**Year 3: Curriculum content coverage**      vi

**Year 3/P4 Half-Termly Tests**

    Autumn Half Term 1:
        Arithmetic      1
        Reasoning      5

    Autumn Half Term 2:
        Arithmetic      9
        Reasoning      13

    Spring Half Term 1:
        Arithmetic      17
        Reasoning      22

    Spring Half Term 2:
        Arithmetic      27
        Reasoning      32

    Summer Half Term 1:
        Arithmetic      37
        Reasoning      42

    Summer Half Term 2:
        Arithmetic      48
        Reasoning      53

**Mark schemes**      60

**Content domain references**      78

**Record sheet**      81

# How to use this book

## Introduction

*Collins Maths Progress Tests for White Rose* have been designed to give you a consistent whole-school approach to teaching and assessing mathematics. Each photocopiable book covers the required mathematics objectives from the 2014 Primary English National Curriculum. For teachers in Scotland, the books can offer guidance and structure that is not provided in the Curriculum for Excellence Experiences and Outcomes or Benchmarks for Numeracy and Mathematics.

As stand-alone tests, the *Collins Maths Progress Tests for White Rose* provide a structured way to assess progress in arithmetic and reasoning skills, to help you identify areas for development, and to provide evidence towards expectations for each year group. Whilst the tests are independent of any textbook-based teaching and learning scheme to allow for maximum flexibility, the content for each test has been selected based on the suggested teaching order in the *White Rose Maths Schemes of Learning*, which are designed to support a mastery approach to teaching and learning.

## Assessment of mathematical skills

At the end of KS1 and KS2, children sit tests to assess the standards they have reached in mathematics. This is done through national curriculum tests (SATs) in Arithmetic and Mathematical Reasoning. *Collins Maths Progress Tests for White Rose* have been designed to provide children with opportunities to explore a range of question types whilst building familiarity with the format, language and style of the SATs.

The Arithmetic tests comprise constructed response questions, presented as context-free calculations, to assess pupils' confidence with a range of mathematics operations as appropriate to the year group. Questions come from the Number, Ratio and Algebra domains.

The Reasoning tests present mathematical problems in a wide range of formats to ensure pupils can fully demonstrate mathematical fluency, mathematical problem solving and mathematical reasoning. They include both selected response questions (e.g. multiple choice, matching, yes/no) and constructed response questions. Questions may draw on all content domains and approximately half of the questions in the Reasoning tests are presented in context.

The tests follow the structure and format of SATs mathematics papers and are pitched at a level appropriate to age-related expectations for the year group. They provide increasing challenge within each year group and across the school, both in terms of content and cognitive demand, but also with increasing numbers of questions to build stamina and resilience. Using the progress tests with your classes at the end of each half-term should help pupils to develop and practise the necessary skills required to complete the national tests with confidence, as well as offering you a snapshot of their progress at those points throughout the year. You can use the results formatively to help identify gaps in knowledge and next teaching steps.

## How to use this book

In this book, you will find twelve photocopiable tests: one arithmetic test and one reasoning test for use at the end of each half term of teaching. Each child will need a copy of the test. You will find Curriculum Content Coverage on page vi indicating the White Rose Scheme of Learning Block and associated Content Domain topics covered in each test across the year group. The specific Content Domain references indicating the year, strand and substrand, e.g. 2N1, for the questions in each test are in the tables on page 78. You may find it useful to make a photocopy of these tables for each child and highlight questions answered incorrectly to help identify any consistent areas of difficulty.

The number of marks available and suggested timing to be allowed are indicated for each test. The number of marks/questions in each test and the length of time allowed increases gradually across the year as summarised in the table below. Note that the Year 2 and Year 6 Summer term tests have been written as full practice papers assuming that all content will have been taught by this point. They mirror the number of marks and time allowed in the end of Key Stage 1 and end of Key Stage 2 test papers.

| Year group | Test | Time allowed | Number of marks |
|---|---|---|---|
| 3 | Autumn 1 Arithmetic | 18 minutes | 20 |
| 3 | Autumn 1 Reasoning | 20 minutes | 15 |
| 3 | Autumn 2 Arithmetic | 18 minutes | 20 |
| 3 | Autumn 2 Reasoning | 20 minutes | 15 |
| 3 | Spring 1 Arithmetic | 22 minutes | 25 |
| 3 | Spring 1 Reasoning | 25 minutes | 20 |
| 3 | Spring 2 Arithmetic | 22 minutes | 25 |
| 3 | Spring 2 Reasoning | 25 minutes | 20 |
| 3 | Summer 1 Arithmetic | 25 minutes | 30 |
| 3 | Summer 1 Reasoning | 30 minutes | 25 |
| 3 | Summer 2 Arithmetic | 25 minutes | 30 |
| 3 | Summer 2 Reasoning | 30 minutes | 25 |

To help you mark the tests, you will find mark schemes at the back of the book. These include the answer requirement, number of marks to be awarded, additional guidance on answers that should or should not be accepted and when to award marks for working in multi-mark questions.

# Test demand

The tests have been written to assess progress in children's arithmetic and mathematical reasoning skills with the content and cognitive demand of questions increasing within each book and across the series to build towards to end of key stage expectations of the SATs. Since the national tests may cover content from the whole key stage, each progress test contains some questions which draw on content from earlier terms or previous year objectives (particularly in autumn term tests). This ensures that prior content and skills are revisited.

The level of demand for each question has been provided within the mark schemes for each test using the notation T (working towards), E (expected standard) or G (greater depth). These ratings are given as an indication of the level of complexity of each question taking into account the thinking skills required to understand what is being asked, the computational complexity in calculating the answer, spatial reasoning or data interpretation required and the response strategy for the question.

# Performance thresholds

The table below provides guidance for assessing how children perform in the tests. Most children should achieve scores at or above the expected standard, with some children working at greater depth and exceeding expectations for their year group. While the thresholds bands do not represent standardised scores, as in the end of key stage SATs, they will give an indication of how pupils are performing against the expected standards for their year group. The thresholds have been set broadly assuming that pupils who achieve greater than 60% will be working at the expected standard and those who score more than 80% are likely to be working at greater depth. However, pupils will all have individual strengths and weaknesses, so it is possible that they could be working towards the expected standard in some areas but at greater depth in others. For this reason, using the content domain coverage tables to identify common areas of difficulty alongside your own professional judgement, will enable you to identify pupils' specific gaps in knowledge and areas where further teaching may be required.

# Tracking progress

A record sheet is provided to help you illustrate to children the areas in which their arithmetic and reasoning skills are strong and where they need to develop. A spreadsheet tracker is also provided via collins.co.uk/assessment/downloads which enables you to identify whole-class patterns of attainment. This can be used to inform your next teaching and learning steps.

# Editable download

All the files are available online in Word and PDF format. Go to collins.co.uk/assessment/downloads to find instructions on how to download. The files are password protected and the password clue is included on the website. You will need to use the clue to locate the password in your book.

You can use these editable files to help you meet the specific needs of your class, whether that be by increasing or decreasing the challenge, by reducing the number of questions, by providing more space for answers or increasing the size of text for specific children.

| Year group | Test | Working towards (T) | Expected standard (E) | Greater depth (G) |
|---|---|---|---|---|
| 3 | Autumn 1 Arithmetic | 11 marks or below | 12–15 marks | 16–20 marks |
| 3 | Autumn 1 Reasoning | 8 marks or below | 9–11 marks | 12–15 marks |
| 3 | Autumn 2 Arithmetic | 11 marks or below | 12–15 marks | 16–20 marks |
| 3 | Autumn 2 Reasoning | 8 marks or below | 9–11 marks | 12–15 marks |
| 3 | Spring 1 Arithmetic | 14 marks or below | 15–19 marks | 20–25 marks |
| 3 | Spring 1 Reasoning | 11 marks or below | 12–15 marks | 16–20 marks |
| 3 | Spring 2 Arithmetic | 14 marks or below | 15–19 marks | 20–25 marks |
| 3 | Spring 2 Reasoning | 11 marks or below | 12–15 marks | 16–20 marks |
| 3 | Summer 1 Arithmetic | 17 marks or below | 18–23 marks | 24–30 marks |
| 3 | Summer 1 Reasoning | 14 marks or below | 15–19 marks | 20–25 marks |
| 3 | Summer 2 Arithmetic | 17 marks or below | 18–23 marks | 24–30 marks |
| 3 | Summer 2 Reasoning | 14 marks or below | 15–19 marks | 20–25 marks |

# Curriculum content coverage

All content objectives from the Year 3 National Curriculum Programme of Study for Mathematics are covered within one or more of the half-termly progress tests across the year. The content for each test is based on the suggested teaching order of the White Rose Maths Schemes of Learning. The table below shows from which teaching blocks the content for each test is drawn. Where the White Rose Maths blocks are devoted to skills or consolidation rather than introduction of new content, these blocks are not covered by the tests. The Summer tests for Year 3 draw on content from previous blocks.

| White Rose Schemes of Learning blocks ||| Collins Maths Progress Tests for White Rose |||||||||||
| Blocks | Weeks | Topics | Autumn 1: Arithmetic | Autumn 1: Reasoning | Autumn 2: Arithmetic | Autumn 2: Reasoning | Spring 1: Arithmetic | Spring 1: Reasoning | Spring 2: Arithmetic | Spring 2: Reasoning | Summer 1: Arithmetic | Summer 1: Reasoning | Summer 2: Arithmetic | Summer 2: Reasoning |
| --- | --- | --- | --- | --- | --- | --- | --- | --- | --- | --- | --- | --- | --- | --- |
| Autumn Block 1 | Weeks 1–3 | Number: Place Value | ✔ | ✔ | | | | | | | ✔ | ✔ | ✔ | ✔ |
| Autumn Block 2 | Weeks 4–8 | Number: Addition and Subtraction | ✔ | ✔ | ✔ | ✔ | | | | | ✔ | ✔ | ✔ | ✔ |
| Autumn Block 3 | Weeks 9–11 | Number: Multiplication and Division | | | ✔ | ✔ | | | | | ✔ | ✔ | ✔ | ✔ |
| Autumn Block 4 | Week 12 | Consolidation | | | | | | | | | | | | |
| Spring Block 1 | Weeks 1–3 | Number: Multiplication and Division | | | | | ✔ | ✔ | ✔ | | ✔ | ✔ | ✔ | ✔ |
| Spring Block 2 | Week 4 | Measurement: Money | | | | | | ✔ | | | | ✔ | | ✔ |
| Spring Block 3 | Weeks 5–6 | Statistics | | | | | | ✔ | | | | ✔ | | ✔ |
| Spring Block 4 | Weeks 7–9 | Measurement: Length and Perimeter | | | | | | | | ✔ | | ✔ | | ✔ |
| Spring Block 5 | Weeks 10–11 | Number: Fractions | | | | | | | ✔ | ✔ | ✔ | ✔ | ✔ | ✔ |
| Spring Block 6 | Week 12 | Consolidation | | | | | | | | | | | | |
| Summer Block 1 | Weeks 1–3 | Number: Fractions | | | | | | | | | ✔ | ✔ | ✔ | ✔ |
| Summer Block 2 | Weeks 4–6 | Measurement: Time | | | | | | | | | | ✔ | | ✔ |
| Summer Block 3 | Weeks 7–8 | Geometry: Properties of Shape | | | | | | | | | | | | ✔ |
| Summer Block 4 | Weeks 9–11 | Measurement: Mass and Capacity | | | | | | | | | | | | ✔ |
| Summer Block 5 | Week 12 | Consolidation | | | | | | | | | | | | |

# Year 3 Autumn Half Term 1: Arithmetic     Name _____

| | |
|---|---|
| **1**   34 + 7 = | **2**   15 + 5 = |
| 1 mark | 1 mark |
| **3**   30 − 6 = | **4**   67 + 23 = |
| 1 mark | 1 mark |
| **5**   ☐ + 9 = 15 | **6**   9 = ☐ − 3 |
| 1 mark | 1 mark |

© HarperCollins*Publishers* Ltd 2019    1

# Year 3 Autumn Half Term 1: Arithmetic

Name _____

**7.** 6 + 8 + 9

1 mark

**8.** 454 − 10 =

1 mark

**9.** 627 + 10 =

1 mark

**10.** 500 − 100

1 mark

**11.** 212 + 7 =

1 mark

**12.** 327 + 9 =

1 mark

Year 3 Autumn Half Term 1: Arithmetic      Name _____

**13.** 415 − 6 =

1 mark

**14.** 345 + 43 =

1 mark

**15.** 876 + 31 =

1 mark

**16.** 414 − 23 =

1 mark

**17.** ☐ = 671 + 39

1 mark

**18.** 700 + ☐ = 1200

1 mark

# Year 3 Autumn Half Term 1: Arithmetic

Name _____

**19** 132 − ☐ = 95

1 mark

**20** ☐ − 46 = 595

1 mark

Total marks ………/20

# Year 3 Autumn Half Term 1: Reasoning    Name _____

**1** Write <, > or = in each circle.

72 ◯ 86

69 ◯ 42

*1 mark*

**2** At the supermarket car park, 78 cars are parked on Level 2 and 54 cars are parked on Level 1.

Circle the **total number** of cars that are parked at the supermarket on the two levels.

102    122    132    134

*1 mark*

**3** In the part–whole model, each number is made by adding the two numbers below it. Fill in the missing numbers.

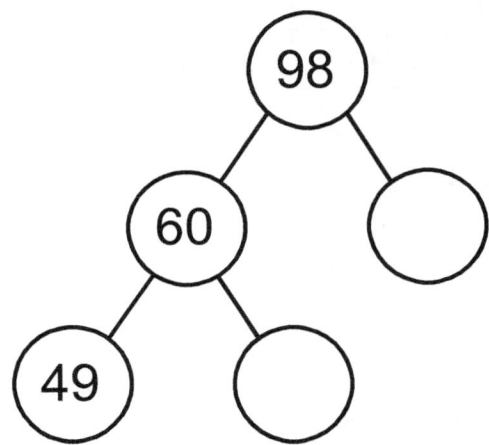

*1 mark*

**4** Draw a line from each sequence to the number that continues it.
One has been done for you.

| 3 | 6 | 9 | 12 | 15 |   | 40 |
|---|---|---|----|----|---|----|
| 90 | 80 | 70 | 60 | 50 |   | 21 |
| 10 | 15 | 20 | 25 | 30 |   | 18 |
| 6 | 9 | 12 | 15 | 18 |   | 35 |

*1 mark*

**Year 3 Autumn Half Term 1: Reasoning**    Name _____

5   Here is an addition.

34 + 54 = 88

**Circle the calculation that can be used to check this addition.**

88 + 54 = 34

88 − 34 = 54

54 − 34 = 88

88 + 34 = 54

1 mark

6   Fill in the missing numbers (in numerals) to complete the table.
The first row has been done for you.

| 10 less | Number in words | 10 more |
|---|---|---|
| 553 | Five hundred and sixty-three | 573 |
|  | Six hundred and one | 611 |
| 499 | Five hundred and nine |  |

2 marks

7   There are **144** packets of crisps in a box.

In this box:
- 56 packets are salt and vinegar.
- 42 packets are sweet chilli.
- The rest of the packets are tomato crisps.

**How many packets of crisps are tomato crisps?**

Show your method

_____ packets of crisps

2 marks

# Year 3 Autumn Half Term 1: Reasoning     Name _____

**8** Mr Jones has 285 spring bulbs to plant in the school grounds.
He has already planted 79 bulbs.

a) How many **more** bulbs does he need to plant?

☐

1 mark

b) Draw place-value counters in the table to show your answer to a).

(100)   (10)   (1)

| Hundreds | Tens | Ones |
|---|---|---|
|  |  |  |

1 mark

**9**  ☐ 234   ☐ 432

Use the numbers on the cards to make this number sentence correct.

☐ > 400 > ☐ < 300

1 mark

# Year 3 Autumn Half Term 1: Reasoning     Name _____

**10** Two trains are travelling to London.
In the train travelling from Manchester there 35 people in the first carriage and 41 people in the second carriage.
In the train travelling from Brighton there are 64 people in the first carriage and 32 people in the second carriage.

How many **more** people are travelling to London from Brighton than from Manchester?

people

2 marks

**11** Ryan says the value of the tens digit in 378 is equal to this number of lollysticks.

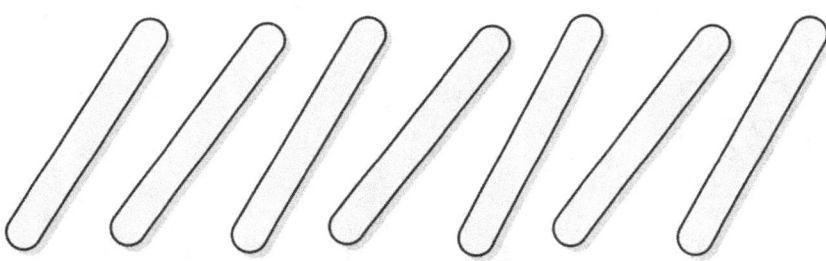

Is Ryan correct?

Circle   Yes   or   No.

Explain how you know.

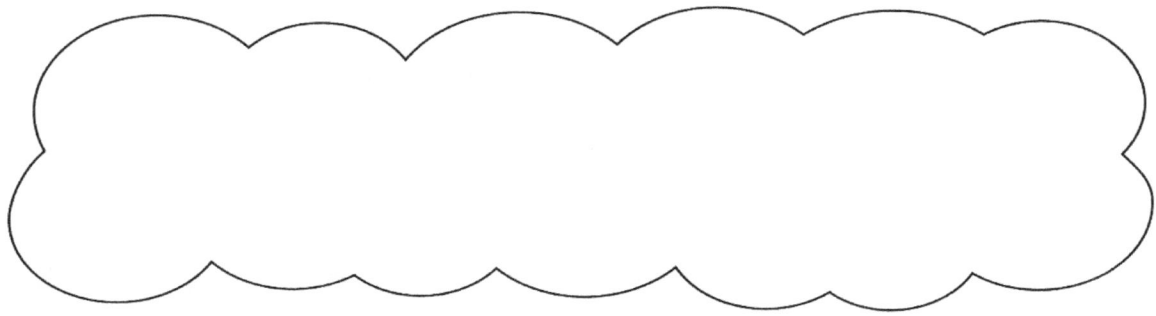

1 mark

Total marks ………/15

# Year 3 Autumn Half Term 2: Arithmetic    Name _____

| 1 | 48 + 9 = |
| 2 | 68 + 36 = |
| 3 | 7 + 9 + 5 = |
| 4 | 5 × ☐ = 50 |
| 5 | ☐ = 10 × 10 |
| 6 | 99 = ☐ + 59 |

Each question: 1 mark

# Year 3 Autumn Half Term 2: Arithmetic

Name _____

**7.** 600 + 200 = 400 + ☐

1 mark

**8.** 409 − 10 =

1 mark

**9.** 678 + 51 =

1 mark

**10.** ☐ = 987 − 74

1 mark

**11.** 875 − 87 =

1 mark

**12.** ☐ − 37 = 108

1 mark

Year 3 Autumn Half Term 2: Arithmetic      Name _____

**13.** 3 × ☐ = 6 × 2                                      1 mark

**14.** 354 + 443 =                                         1 mark

**15.** 538 − 216 =                                         1 mark

**16.** 198 + 704 =                                         1 mark

**17.** 628 − 409 =                                         1 mark

**18.** ☐ = 8 × 8                                           1 mark

Year 3 Autumn Half Term 2: Arithmetic       Name _____

**19**  4 × ☐ = 48

1 mark

**20**  56 ÷ 7 =

1 mark

Total marks ………/20

# Year 3 Autumn Half Term 2: Reasoning    Name _____

**1** Tick (✓) the number that is shown by the abacus.

Sixteen ☐

Seven ☐

Sixty-one ☐

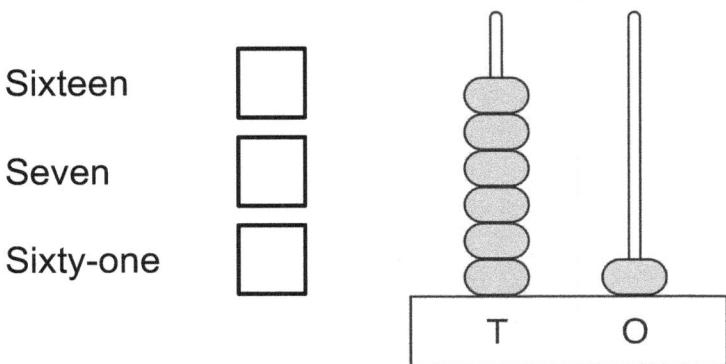

1 mark

**2** Class 3G give one full box of pencils to Class 3L.

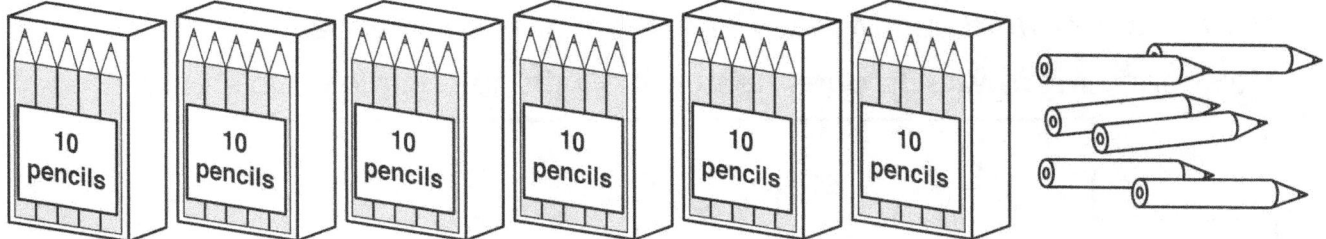

How many pencils do they have left?

☐

1 mark

**3**  + ☐ + △ = 100

The value of a triangle is 50.
The value of a circle is 20.

What is the value of the square?

1 mark

**4** Write True or False next to each statement.

93 < nine tens and six ones ☐

67 = six tens and seventeen ones ☐

31 > one ten and nine ones ☐

99 < one hundred ☐

1 mark

# Year 3 Autumn Half Term 2: Reasoning    Name _____

**5** Fill in the **missing numbers** in this multiplication grid.

| × | 3 | 4 | 8 |
|---|---|---|---|
| 4 | 12 |  |  |
|  | 15 | 20 | 40 |
| 9 | 27 |  | 72 |

2 marks

**6** Joe is writing a **950-word** story.
He wrote 330 words on Monday and 405 words on Tuesday.

**How many more words** does he need to write to complete his story?

Show your method

_____ words

2 marks

**7** Use the numbers on the cards to complete the number tracks.

| 300 | 550 | 900 | 600 |

|  | 350 | 400 | 450 | 500 |  |

| 400 | 500 |  | 700 | 800 |  |

2 marks

# Year 3 Autumn Half Term 2: Reasoning    Name _____

**8** Sunflowers are sold in bunches of 4.
Tia buys 24 sunflowers.

How many bunches does she buy?

 bunches

1 mark

**9** Write four **different** numbers to make this statement correct.

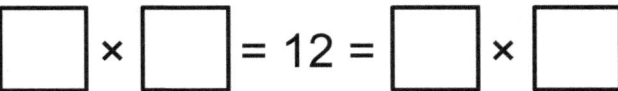

1 mark

**10** Here is a white tile. It is **3 counters** long.

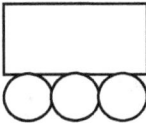

A grey tile is eight times as long.

The tiles are joined together.

Find the total length of the two tiles in counters.

| Show your method | |
|---|---|
| | counters |

2 marks

# Year 3 Autumn Half Term 2: Reasoning     Name _____

**11** Ben says multiples of 100 always have 3 digits.

Is Ben correct?

Circle Yes or No.

Explain how you know.

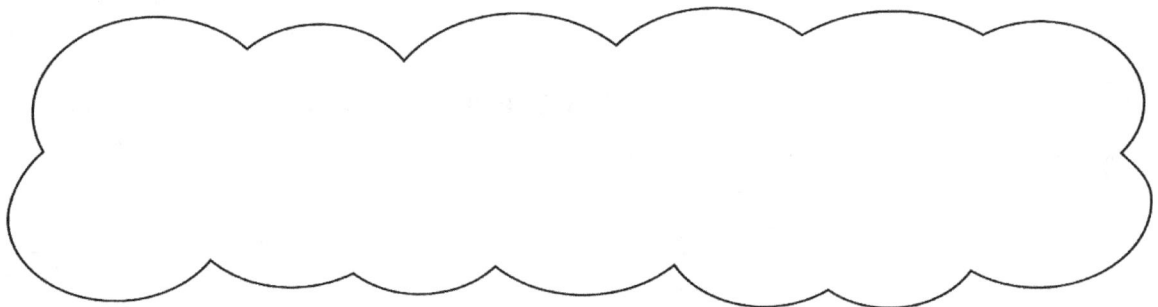

1 mark

Total marks ………/15

# Year 3 Spring Half Term 1: Arithmetic

Name _____

| 1 | 84 + 57 = |
|---|---|

1 mark

| 2 | 5 + 9 + 9 = |
|---|---|

1 mark

| 3 | 43 − ☐ = 4 |
|---|---|

1 mark

| 4 | ☐ = 5 × 12 |
|---|---|

1 mark

| 5 | 34 + ☐ = 100 |
|---|---|

1 mark

| 6 | 354 − 200 = |
|---|---|

1 mark

# Year 3 Spring Half Term 1: Arithmetic

Name _____

**7**   451 + 249 =

*1 mark*

**8**   7 × 8 =

*1 mark*

**9**   782 − 681 =

*1 mark*

**10**   ☐ = 3 × 9

*1 mark*

**11**   434 − ☐ = 334

*1 mark*

**12**   32 ÷ 4 =

*1 mark*

# Year 3 Spring Half Term 1: Arithmetic

**13.** 640 − 10 =

**14.** 354 − ☐ = 326

**15.** ☐ = 21 ÷ 3

**16.** 8 × 9 =

**17.** 567 + 396 =

**18.** 23 × 3

# Year 3 Spring Half Term 1: Arithmetic     Name _____

**19**  ☐ = 42 × 4

1 mark

**20**  45 ÷ 3 =

1 mark

**21**  90 ÷ 6 =

1 mark

**22**  899 + 10 =

1 mark

**23**  35 × 4 =

1 mark

**24**  87 ÷ 3 =

1 mark

# Year 3 Spring Half Term 1: Arithmetic

Name _____

**25**   4 × ☐ = 72

1 mark

Total marks ………/25

# Year 3 Spring Half Term 1: Reasoning

Name _____

**1** Inez makes a **tally chart** of all the pieces of fruit in her fruit bowl.

| Apple | III |
| --- | --- |
| Banana | ЖI |
| Pear | II |

Tick (✓) one box below that shows **all** the fruit in Inez's fruit bowl.

1 mark

**2** Here is a bar model.

| 19 ||
| --- | --- |
| 3 | 16 |

Raj uses this to write an **addition calculation**.

16 + 3 = 19

Write an **inverse** calculation to check 16 + 3 = 19

|  |
| --- |

1 mark

**3** Circle **all** the calculations that match the array.

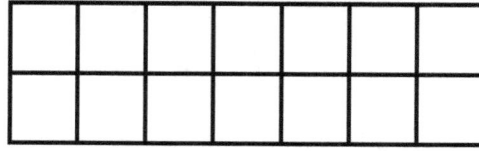

2 × 7     2 − 7     7 + 7     14 ÷ 2

1 mark

# Year 3 Spring Half Term 1: Reasoning    Name _____

**4**  Mya has a birthday party.
 9 boys, 8 girls and 7 adults come to her party.

a) How many people come to her party?

| people |
|---|

1 mark

There were 10 chocolate cupcakes and 10 strawberry cupcakes at the party. 14 cupcakes were eaten.

b) How many cupcakes were left over?

| cupcakes |
|---|

1 mark

**5**  Write these numbers in order, starting with the **largest**.

| 686 | 866 | 868 | 668 | 688 |
|---|---|---|---|---|

| | | | | |
|---|---|---|---|---|

largest                                         smallest

1 mark

**6**  Sam buys this pen.

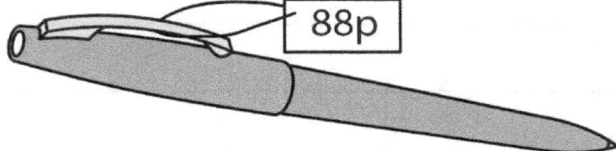

He pays for the pen with a **£2 coin**.

Tick (✓) all the different ways Sam could get his **change**.

10p  2p                          ☐

£1  10p  2p                      ☐

5p  5p  1p  1p                   ☐

50p  50p  20p  1p  1p            ☐

2p 2p 2p 2p 2p 2p                ☐

20p 20p 10p 50p  10p  2p         ☐

2 marks

Year 3 Spring Half Term 1: Reasoning         Name _____

7  Fill in the missing boxes.

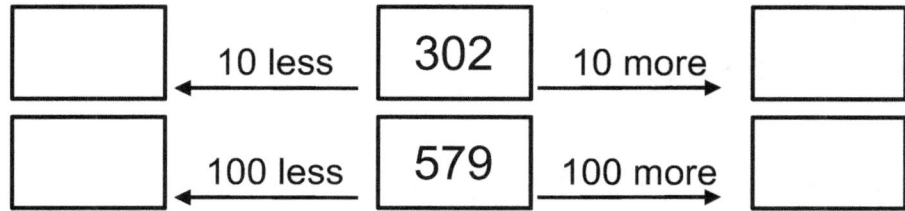

2 marks

8

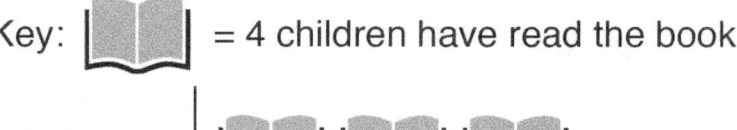

Number of children who have read the book

a) What was the most **popular** book to read?

|  |
|---|

1 mark

b) How many children read 'Jumping Jack'?

| children |
|---|

1 mark

c) How many **more** children read the 'The Magic Pineapple' than 'Frog's Socks'?

| children |
|---|

1 mark

Year 3 Spring Half Term 1: Reasoning    Name _____

**9** Mike buys a new comic book costing **£2 and 15 pence** and one small cookie costing **49 pence**.
He pays with a **£5 note**.

How much **change** does he get?

pence

2 marks

**10** Eve and Raj have **each** built a tower.

Eve says Raj's tower is 2 times taller than her tower.
Raj says his tower is 10 times taller.

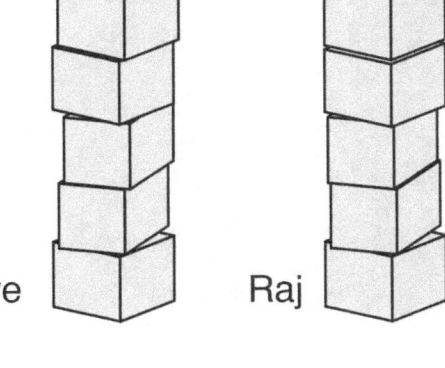

Eve     Raj

Who do you agree with?

Explain your answer.

1 mark

# Year 3 Spring Half Term 1: Reasoning    Name _____

**11** Martin has **8** bags of marbles.
Each of his bags of marbles contains **25** marbles.
Sue has **3** bags of marbles.
Each of her bags of marbles contains **40** marbles.

How many marbles do they have altogether?

Show your method

_____ marbles

2 marks

**12** Fill in the missing numbers to complete this **addition** calculation.

```
   1 5 ▢
+  ▢ ▢ 4
  -------
   7 1 7
   1
```

2 marks

Total marks ………/20

# Year 3 Spring Half Term 2: Arithmetic

Name _____

**1)** 45 + 35 =

*1 mark*

**2)** 10 × ☐ = 70

*1 mark*

**3)** 60 ÷ 5 =

*1 mark*

**4)** 101 − 10 =

*1 mark*

**5)** 272 + 700 =

*1 mark*

**6)** 8 × 6 =

*1 mark*

# Year 3 Spring Half Term 2: Arithmetic

Name _____

**7**   359 + 8 =

1 mark

**8**   443 − 42 =

1 mark

**9**   789 + 151 =

1 mark

**10**   ☐ = 765 − 439

1 mark

**11**   ☐ = 8 × 9

1 mark

**12**   901 − 100 =

1 mark

# Year 3 Spring Half Term 2: Arithmetic    Name _____

| 13 | $4 \div 10 = \frac{\square}{10}$ |
| 14 | $3 \times \boxed{\phantom{00}} = 24$ |
| 15 | $75 \div 5 =$ |
| 16 | $895 + 10 =$ |
| 17 | $38 \times 4 =$ |
| 18 | $8 \div 10 = \frac{\square}{10}$ |

(1 mark each)

# Year 3 Spring Half Term 2: Arithmetic

Name _____

**19** $\frac{1}{3}$ of 21 =

1 mark

**20** $\frac{1}{10}$ of 100 =

1 mark

**21** 800 + ☐ = 1400

1 mark

**22** $\frac{2}{5}$ of 35 =

1 mark

**23** 1 ÷ ☐ = 0.1

1 mark

**24** $\frac{1}{4}$ of ☐ = 5

1 mark

# Year 3 Spring Half Term 2: Arithmetic

Name _____

**25** ☐ = $\frac{5}{8}$ of 96

1 mark

Total marks ........./25

**Year 3 Spring Half Term 2: Reasoning**    Name _____

1  Gina put some counters into groups to solve a calculation.

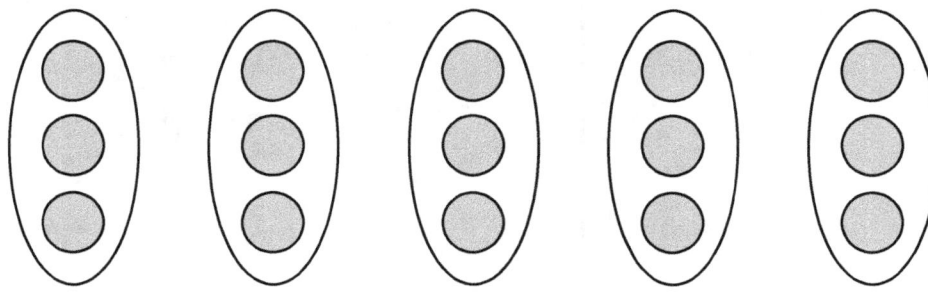

Tick (✓) the calculation Gina solves.

15 ÷ 3 = ☐

15 ÷ 5 = ☐

5 + 15 = ☐

1 mark

2  Sophie takes 10 books out of the cupboard.
4 books are fiction books.
6 books are non-fiction books.

What fraction of the books are **fiction** books?

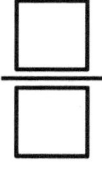

1 mark

3  Match the number in **words** to the number in **numerals**.

Nineteen            9

Twenty-nine         92

Ninety-two          19

Nine                29

1 mark

Year 3 Spring Half Term 2: Reasoning        Name _____

**4** The numbers in the two triangles **add up** to the number in the circle.

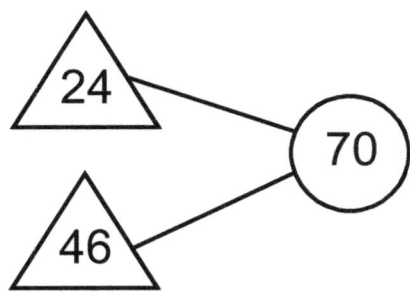

Fill in the missing numbers.

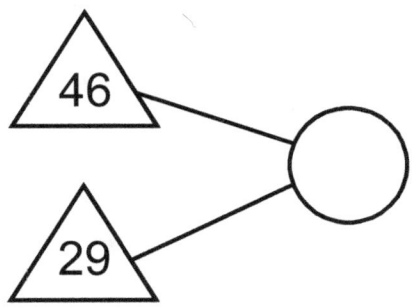  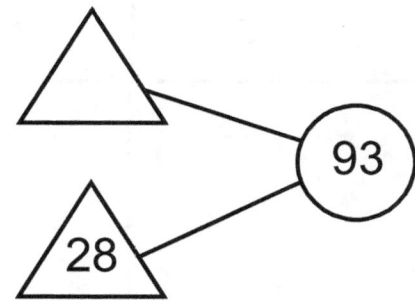

2 marks

**5** Measure the **length** of this line, in centimetres.

_____

| cm |

1 mark

**6** A group of **32** children each chose their favourite sport.

$\frac{1}{4}$ liked football.

$\frac{1}{8}$ liked rugby.

The rest of the children liked swimming.

**How many** children liked swimming?

2 marks

Year 3 Spring Half Term 2: Reasoning         Name _____

7  In this diagram, you **multiply** the numbers next to each other and write the answer in the block **above**.

Fill in the missing numbers.

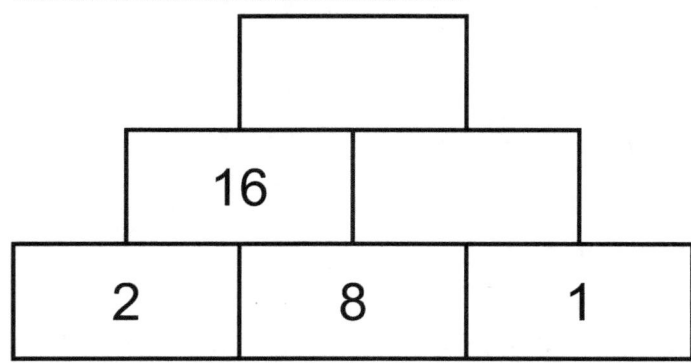

2 marks

8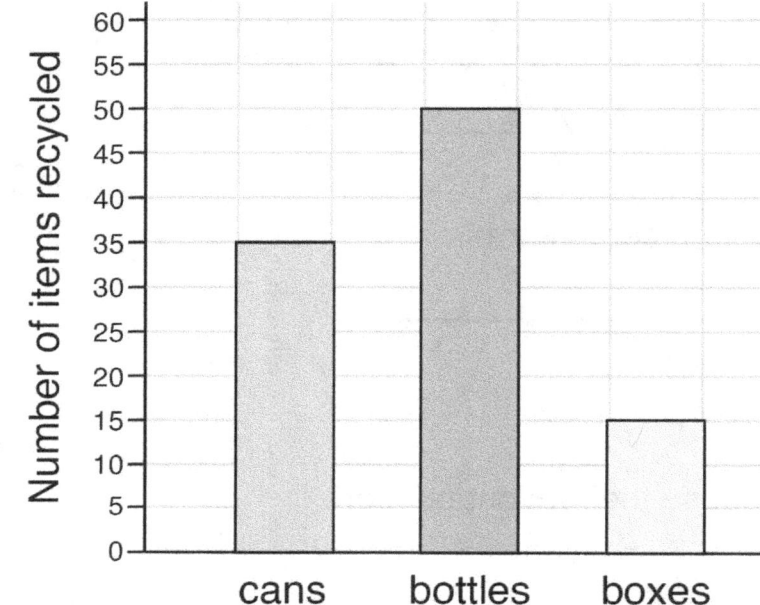

a) How many **bottles** were recycled?

1 mark

b) How many **fewer** boxes than bottles were recycled?

1 mark

c) What is the total number of **cans and bottles** recycled?

1 mark

# Year 3 Spring Half Term 2: Reasoning     Name _____

**9** Here is a table showing the heights of **four** children.

| Child | Height |
|---|---|
| Ben | 1 m 5 cm |
| Tia | 109 cm |
| Raj | 1 m 15 cm |
| Sam | 1 m 70 mm |

Order the children's heights. Start with the **shortest**.

| | | | |
|---|---|---|---|

shortest                                                                                    tallest

2 marks

**10**

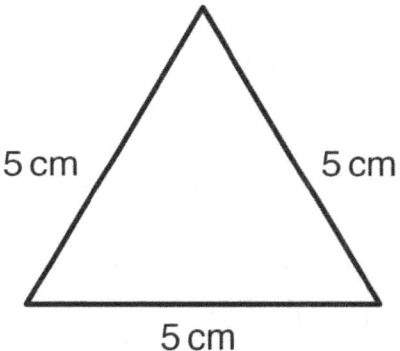

Eva says the perimeter of the triangle is 25 cm.

Is Eva correct?

Circle   Yes   or   No.

Explain how you know.

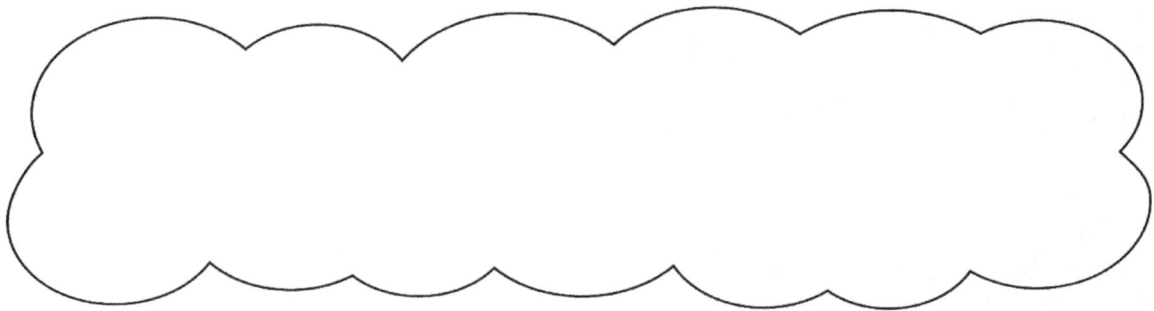

1 mark

# Year 3 Spring Half Term 2: Reasoning      Name _____

**11** Grandma Meg has 12 m of wool left in her ball.
She used 3 m to knit a scarf and 108 cm to knit a pair of mittens.
**How much wool does she have left?**

Show your method

cm

2 marks

**12** Write the missing numbers to complete the fraction statements.

$\frac{1}{10}$ of 50 = $\frac{1}{3}$ of ☐

$\frac{1}{☐}$ of 24 = $\frac{2}{5}$ of 30

2 marks

Total marks ………/20

# Year 3 Summer Half Term 1: Arithmetic

Name _____

**1.** 68 + 54 =

*1 mark*

**2.** 8 + 7 + 6 =

*1 mark*

**3.** 300 + 400 =

*1 mark*

**4.** 905 − 100 =

*1 mark*

**5.** 195 + 10 =

*1 mark*

**6.** ☐ = 375 − 124

*1 mark*

# Year 3 Summer Half Term 1: Arithmetic

Name _____

**7**   4 × 7 =

1 mark

**8**   55 ÷ 11 ☐

1 mark

**9**   345 + 55 =

1 mark

**10**   ☐ = 6 × 4

1 mark

**11**   567 + 245 =

1 mark

**12**   ☐ + 899 = 999

1 mark

# Year 3 Summer Half Term 1: Arithmetic

Name _____

**13** 82 × 3 =

1 mark

**14** 5 ÷ 10 = $\frac{\phantom{0}}{10}$

1 mark

**15** ☐ = 48 × 3

1 mark

**16** $\frac{1}{10}$ of 90 =

1 mark

**17** 702 − 117 =

1 mark

**18** $\frac{5}{7} + \frac{2}{7} =$

1 mark

# Year 3 Summer Half Term 1: Arithmetic

**19** 92 ÷ 4 =

1 mark

**20** $\frac{5}{6} - \frac{2}{6} =$

1 mark

**21** 305 − 10 =

1 mark

**22** 674 − ☐ = 452

1 mark

**23** $\frac{2}{3}$ of 12 =

1 mark

**24** 8 × ☐ = 96

1 mark

# Year 3 Summer Half Term 1: Arithmetic     Name _____

**25.** 209 − 10 =                                1 mark

**26.** ☐ ÷ 10 = $\frac{2}{10}$                   1 mark

**27.** $\frac{8}{9}$ − ☐ = $\frac{5}{9}$          1 mark

**28.** $\frac{1}{8}$ + $\frac{2}{8}$ + $\frac{3}{8}$ =   1 mark

**29.** ☐ = $\frac{3}{4}$ of 60                    1 mark

**30.** 3 × ☐ = 84                                 1 mark

Total marks ………/30

**Year 3 Summer Half Term 1: Reasoning**   Name _____

1  Tick (✓) two shapes that have exactly $\frac{2}{4}$ shaded.

[bar divided in 2, right half shaded]

[bar divided in 4, first and last quarters shaded]

[bar divided in 4, first two quarters shaded]

[bar divided in 4, third quarter shaded]

1 mark

2  Paolo has **21** counters.
He put them into **threes**, like this:

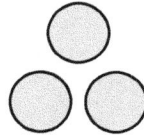

How many **groups of three** can he make?

1 mark

3  Ken buys a bouncy ball for 75p.
Tick (✓) **all** the different ways Ken can pay for the ball

50p   2p   2p   2p   ☐

10p   10p   20p   20p   10p   5p   ☐

50p   20p   1p   1p   1p   ☐

10p   10p   10p   10p   10p   10p   10p   5p   ☐

20p   50p   2p   2p   1p   ☐

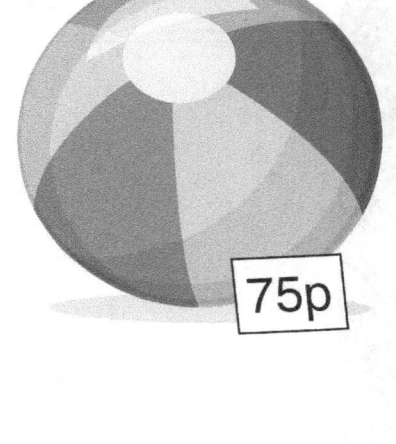

1 mark

# Year 3 Summer Half Term 1: Reasoning    Name _____

**4** The numbers in each row and in each column add up to 100.

Fill in the missing numbers.

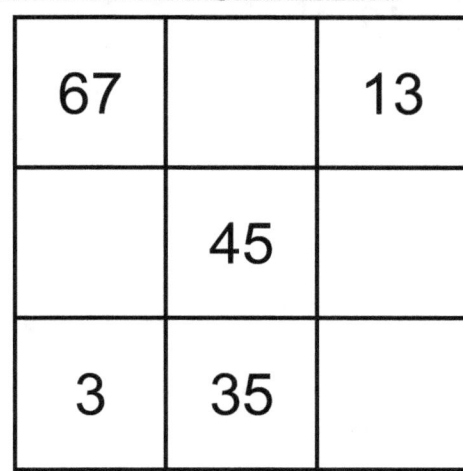

2 marks

**5** Order this set of numbers from **largest to smallest**.

$\frac{2}{8}$    1    $\frac{7}{8}$    $\frac{4}{8}$

☐ ☐ ☐ ☐

Largest            Smallest

1 mark

**6** Here is a **fraction wall** diagram.

Use the diagram to help you fill in the missing numbers.

$$\frac{1}{5} = \frac{\square}{10}$$

$$\frac{8}{\square} = \frac{\square}{5}$$

2 marks

Year 3 Summer Half Term 1: Reasoning    Name _____

7  Use the numbers on the cards to complete the statements.

There are ☐ hours in a day.

There are ☐ months in a year.

There are ☐ minutes in an hour.

There are ☐ days in a year.

There are ☐ days in a leap year.

2 marks

8  Write the time shown on each clock.

a) In the morning, Ben reads a book.

Write the time in words.

☐

1 mark

b) In the evening, Ben watches TV.
   Write the time as a 12-hour digital time.

☐ : ☐

1 mark

c) Ben goes to bed at this time.
   Write the time in words.

21:15

☐

1 mark

# Year 3 Summer Half Term 1: Reasoning     Name _____

**9** At the start of the week there are 235 T-shirts in the factory.
During the week:
- 147 more T-shirts are made
- 151 T-shirts are sold.

How many T-shirts are left in the factory at the end of the week?

T-shirts

2 marks

**10 a)** Toby and Jack bake some biscuits. It takes:
- 20 minutes to make the mixture
- 10 minutes to cut the biscuits
- 12 minutes to bake the biscuits.

They start baking at **ten past three**.

Write the time when the biscuits will be ready.

1 mark

**b)** Toby spends 1 minute 15 seconds icing one biscuit.
Jack spends 80 seconds icing one biscuit.

How many **more** seconds does Jack spend icing a biscuit than Toby?

1 mark

**Year 3 Summer Half Term 1: Reasoning**          Name _____

**11** Jen and Raj share a large pizza.
The pizza has been cut into 12 slices.
Jen eats 4 slices and Raj eats **1 less** slice than Jen.

What fraction of the pizza is **left over**?

slices

2 marks

**12** Write this number in **words**.

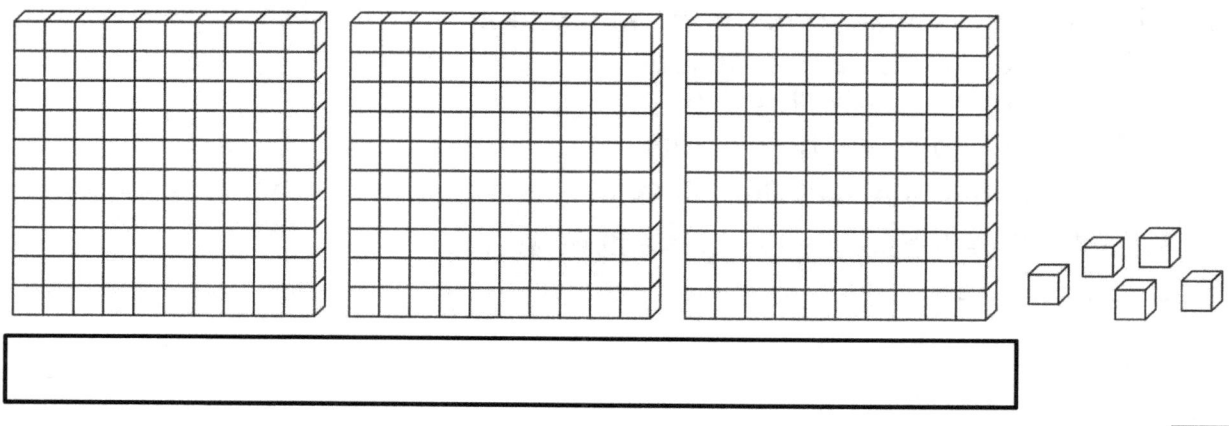

1 mark

**13** Kay has **96** bulbs. She plants **4** bulbs in each flower pot.
Rob has **78** bulbs. He plants **3** bulbs in each flower pot.

How many **more** flower pots does Rob need than Kay?

flower pots

2 marks

Year 3 Summer Half Term 1: Reasoning        Name _____

14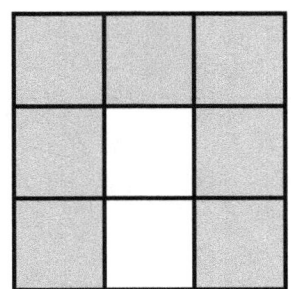

a) Write an addition fraction calculation to match this diagram.

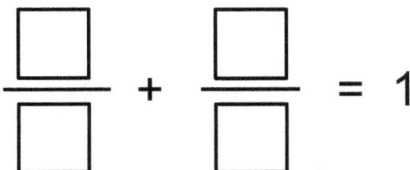

1 mark

b) Write a subtraction fraction calculation to match this diagram.

$1 - \dfrac{\square}{\square} = \dfrac{\square}{\square}$

1 mark

15 Would you rather have $\dfrac{1}{3}$ of £45 or $\dfrac{3}{8}$ of £24?

Explain your answer.

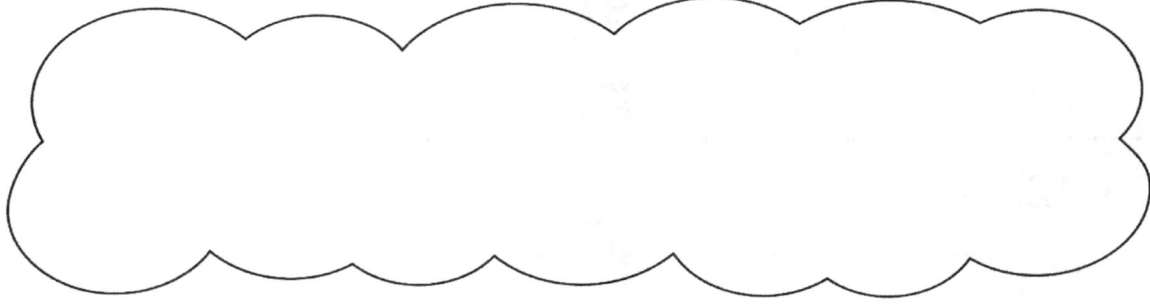

1 mark

Total marks ………/25

# Year 3 Summer Half Term 2: Arithmetic          Name _____

**1)** 375 + 10 =

*1 mark*

**2)** 909 − 10 =

*1 mark*

**3)** 137 + 44 =

*1 mark*

**4)** 4 × 4 =

*1 mark*

**5)** 125 + 125 =

*1 mark*

**6)** 683 − 21 =

*1 mark*

# Year 3 Summer Half Term 2: Arithmetic     Name _____

| 7 | 609 − 129 |
|---|---|

| 8 | ☐ = 11 × 8 |
|---|---|

| 9 | $6 \div 10 = \dfrac{\phantom{0}}{10}$ |
|---|---|

| 10 | 32 ÷ 4 = |
|---|---|

| 11 | ☐ = 496 + 387 |
|---|---|

| 12 | $\dfrac{6}{8} + \dfrac{1}{8} =$ |
|---|---|

1 mark (each)

# Year 3 Summer Half Term 2: Arithmetic     Name _____

**13**  63 × 3 =

1 mark

**14**  ☐ = 903 − 482

1 mark

**15**  54 × 8 =

1 mark

**16**  $\frac{1}{8}$ of 64 =

1 mark

**17**  75 ÷ 3 =

1 mark

**18**  12 × ☐ = 48

1 mark

Year 3 Summer Half Term 2: Arithmetic    Name _____

19) $9 \div 10 = \dfrac{\phantom{0}}{10}$

1 mark

20) $\boxed{\phantom{00}} + 709 = 909$

1 mark

21) $678 + 495 =$

1 mark

22) $503 - 298 =$

1 mark

23) $\dfrac{3}{10}$ of $120 =$

1 mark

24) $\boxed{\phantom{00}} = 56 \div 4$

1 mark

# Year 3 Summer Half Term 2: Arithmetic     Name _____

**25)** 48 ÷ ☐ = 8                    **1 mark**

**26)** 999 + 100 =                   **1 mark**

**27)** $\frac{1}{6} + \frac{2}{6} + \frac{1}{6} =$      **1 mark**

**28)** ☐ ÷ 10 = 0.1                  **1 mark**

**29)** $\frac{1}{8}$ of ☐ = 5         **1 mark**

**30)** $\frac{2}{10} + \frac{1}{10} + $ ☐ $= \frac{9}{10}$   **1 mark**

Total marks ………/30

# Year 3 Summer Half Term 2: Reasoning          Name _____

1  Fill in the missing numbers on the addition grid.

| +  | 16 |    | 68 |
|----|----|----|----|
| 25 |    | 70 |    |
|    | 43 | 72 | 95 |

2 marks

2  30 blue pens are put into packs of 10.
   15 red pens are put into packs of 5.

   **How many packs are there altogether?**

   Show your method

                                                                packs

2 marks

3  Match each measuring tool to the correct unit of measure.

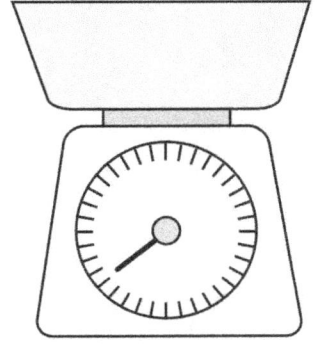 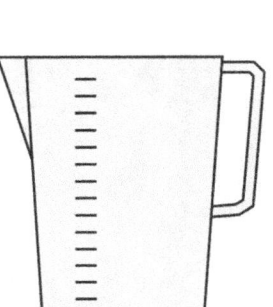 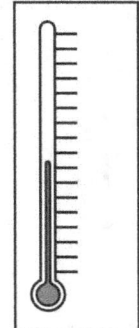 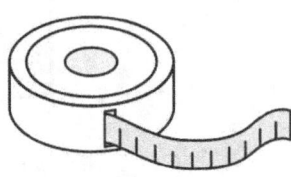

   | °C |    | kg |    | cm |    | ml |

1 mark

# Year 3 Summer Half Term 2: Reasoning    Name _____

**4** A block of butter weighs **350 g**.

Draw an arrow on the dial to show the mass of the butter.

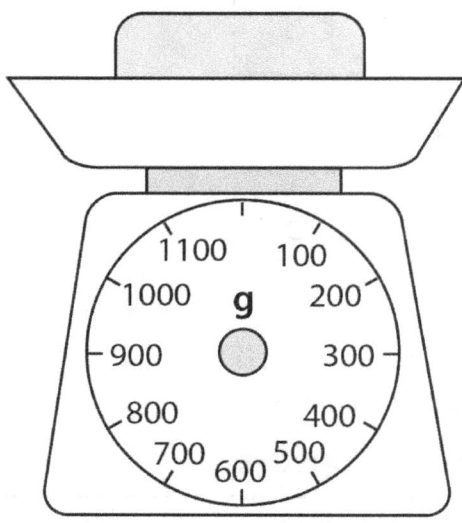

1 mark

**5** Write the missing numbers in the fraction sequence.

$\dfrac{\square}{10}$    $\dfrac{4}{10}$    $\dfrac{5}{10}$    $\dfrac{6}{10}$    $\dfrac{\square}{10}$    $\dfrac{8}{10}$    $\dfrac{9}{10}$    $1$    $1\dfrac{\square}{10}$

1 mark

**6** Jack says: "This 3D shape is a cuboid."

Do you agree?

Explain your reasoning.

1 mark

# Year 3 Summer Half Term 2: Reasoning    Name _____

7   Here are three bottles of milk.

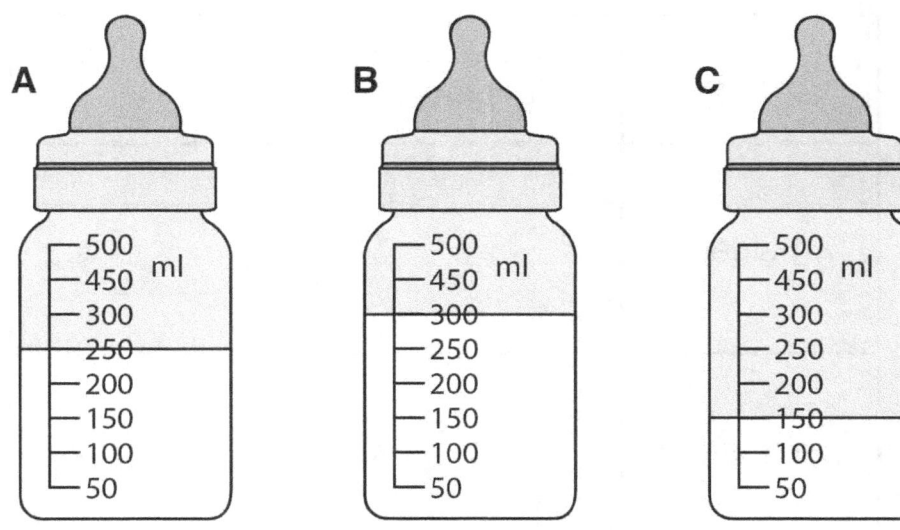

a) What is the **difference** between the greatest and the smallest amounts of milk?

[        ] ml

1 mark

b) Calculate the total amount of milk in **all** three bottles.

[        ] ml

1 mark

8   At the airport, there **678** suitcases that need to be put on different planes.
    **234** suitcases are for the flight to Paris.
    **128** suitcases are for the flight to Berlin.
    The **remaining** suitcases are for the flight to Palma.

    How **many** suitcases need to be put on the plane to Palma?

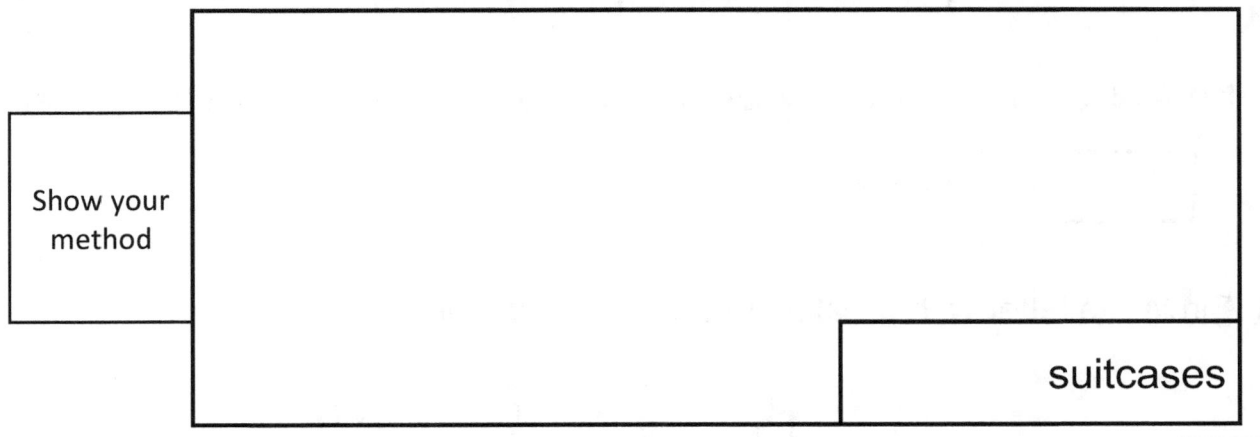

2 marks

# Year 3 Summer Half Term 2: Reasoning    Name _____

**9** Draw 2D shapes to complete the Carroll diagram.

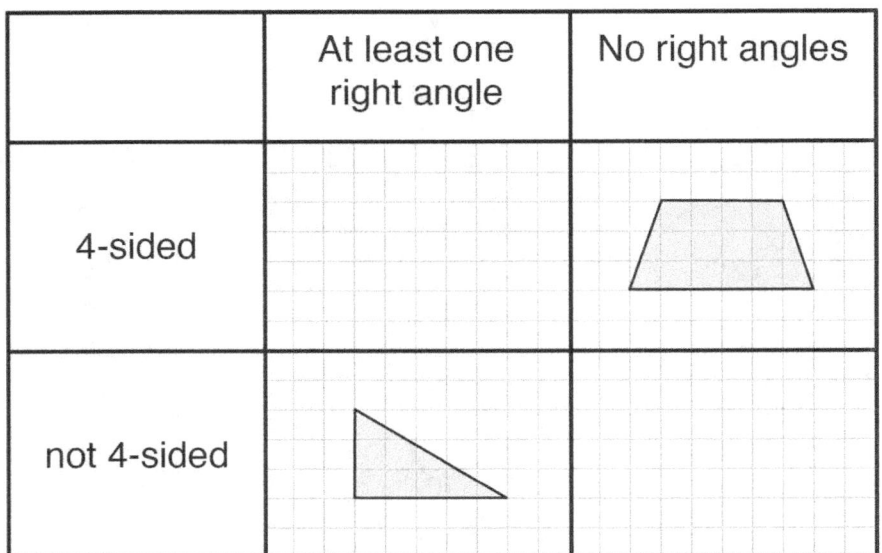

2 marks

**10** Here are some parcels.

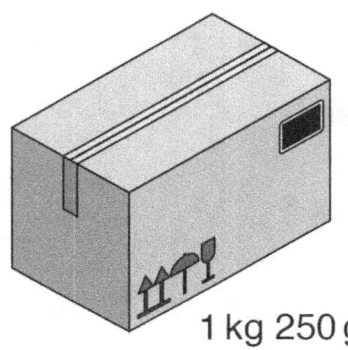
1 kg 250 g

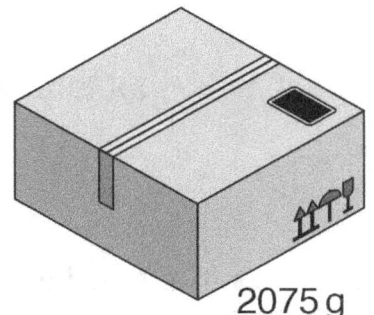
2075 g

2½ kg

a) Order the parcels from heaviest to lightest.

☐   ☐   ☐

1 mark

b) What is the difference in mass between the heaviest and lightest parcels?

☐

1 mark

**11** Circle the letter with parallel **and** perpendicular lines.

J    K    O    F    M

1 mark

Year 3 Summer Half Term 2: Reasoning    Name _____

**12 a)** Draw the number 409 on the place-value chart.

| Hundreds | Tens | Ones |
|---|---|---|
|  |  |  |

1 mark

**b)** Write the symbols < or > to make the statement correct.

1 mark

**13** Jared is facing towards the arrow.

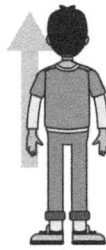

He makes a turn, so he is now in this new position.

Jared says he has done a quarter turn **anticlockwise**.
Hannah says he has done a three-quarter turn **clockwise**.

Who do you agree with?

Explain your answer.

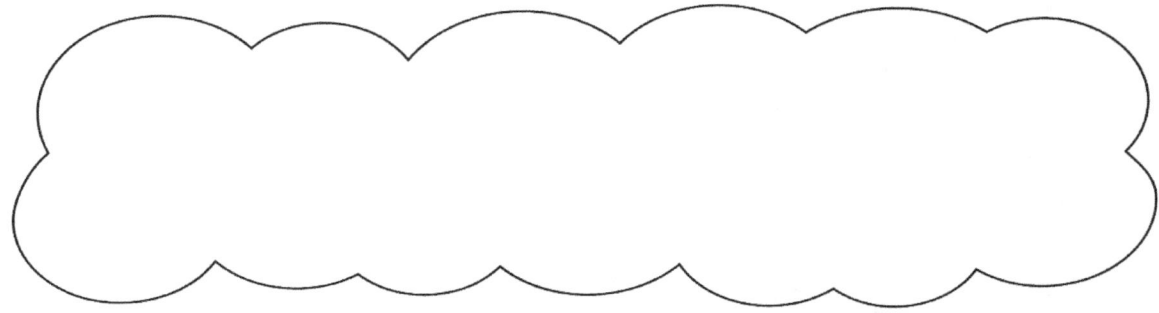

1 mark

**14** Suki is at the post office.
Her small parcel weighs the **same** as 1 large letter and 3 small letters.
The small parcel weighs 196 g.
The large letter weighs 100 g.

What does **each** small letter weigh?

g

2 marks

**15** Use the numbers on the digit cards to write the calculation with the answer closest to 100.

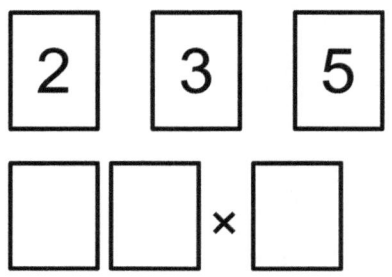

☐☐ × ☐

1 mark

**16** The chart shows the favourite sports of Year 3 girls and boys.

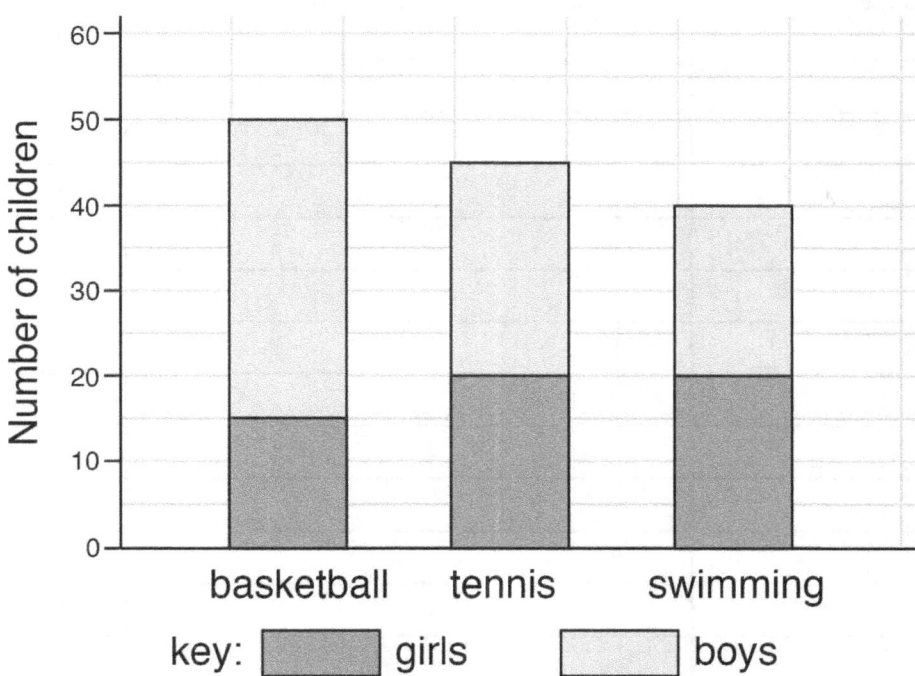

a) Altogether, how many girls like basketball and tennis?

[         girls         ]

1 mark

b) How many **more** boys chose basketball than chose swimming?

[         boys         ]

1 mark

Total marks ......... /25

# Year 3 Autumn Half Term 1: Arithmetic Mark Scheme

| Question | Requirement | Mark | Additional guidance | Level of demand |
|---|---|---|---|---|
| 1 | 41 | 1 | | T |
| 2 | 20 | 1 | | T |
| 3 | 24 | 1 | | T |
| 4 | 90 | 1 | | T |
| 5 | 6 | 1 | | T |
| 6 | 12 | 1 | | T |
| 7 | 23 | 1 | | T |
| 8 | 444 | 1 | | T |
| 9 | 637 | 1 | | E |
| 10 | 400 | 1 | | E |
| 11 | 219 | 1 | | E |
| 12 | 336 | 1 | | E |
| 13 | 409 | 1 | | E |
| 14 | 388 | 1 | | E |
| 15 | 907 | 1 | | E |
| 16 | 391 | 1 | | E |
| 17 | 710 | 1 | | G |
| 18 | 500 | 1 | | G |
| 19 | 37 | 1 | | G |
| 20 | 641 | 1 | | G |

**Threshold scores**
Working towards the expected standard (T): 11 or fewer
Working at the expected standard (E): 12–15
Working at greater depth (G): 16–20

**Balance of difficulty of questions in the paper**
8 marks at working towards (T)
8 marks at the expected standard (E)
4 marks at working at greater depth (G)

# Year 3 Autumn Half Term 1: Reasoning Mark Scheme

| Question | Requirement | Mark | Additional guidance | Level of demand |
|---|---|---|---|---|
| 1 | <br>> | 1 | | T |
| 2 | 132 | 1 | Accept alternative unambiguous positive indications, e.g. 132 ticked or underlined. | T |
| 3 | 38 and 11 | 1 | Both correct missing numbers must be given for the award of **1 mark**. | T |
| 4 | 3  6  9  12  15 — 40<br>90  80  70  60  50 — 21<br>10  15  20  25  30 — 18<br>6  9  12  15  18 — 35 | 1 | All numbers must be matched to the correct sequence for the award of **1 mark**. | T |
| 5 | 88 − 34 = 54 | 1 | Accept alternative unambiguous positive indications, e.g. 88 − 34 = 54 ticked or underlined. | E |
| 6 | 10 less / 10 more<br>553  Five hundred and sixty-three  573<br>591  Six hundred and one  611<br>499  Five hundred and nine  519 | 2 | Award **1 mark** for each correct answer. | E |
| 7 | 46 | 2 | Award **2 marks** for the correct answer of 46.<br>If the answer is incorrect, award **1 mark** for evidence of an appropriate method, e.g.<br>144 − 56 − 42 =<br>Or<br>56 + 42 = 98<br>144 − 98 = | E |
| 8a | 206 | 1 | | E |
| 8b | Hundreds: 100, 100; Tens: (none); Ones: 1, 1, 1, 1, 1, 1 | 1 | Do **not** accept 2 and 6 written as numerals in the chart. Accept a wrong answer, if it correctly follows on from an incorrect answer to the first part. | E |
| 9 | **432** > 400 > **234** < 300 | 1 | | E |
| 10 | 20 | 2 | Award **2 marks** for the correct answer of 20.<br>If the answer is incorrect, award **1 mark** for evidence of an appropriate method, e.g.<br>35 + 41 = 76<br>64 + 32 = 96<br>96 − 76 = 20 | G |

# Year 3 Autumn Half Term 1: Reasoning Mark Scheme

| 11 | No; the number of lollysticks given is seven however the tens digit is referring to the number seventy. | 1 | Do **not** accept vague or incomplete explanations. Do **not** accept 'No' circled without a correct explanation. Accept 'Yes' or 'No' circled with the correct explanation. | G |

**Threshold scores**
Working towards the expected standard (T): 8 or fewer
Working at the expected standard (E): 9–11
Working at greater depth (G): 12–15

**Balance of difficulty of questions in the paper**
4 marks at working towards (T)
8 marks at the expected standard (E)
3 marks at working at greater depth (G)

# Year 3 Autumn Half Term 2: Arithmetic Mark Scheme

| Question | Requirement | Mark | Additional guidance | Level of demand |
|---|---|---|---|---|
| 1 | 57 | 1 | | T |
| 2 | 104 | 1 | | T |
| 3 | 21 | 1 | | T |
| 4 | 10 | 1 | | T |
| 5 | 100 | 1 | | T |
| 6 | 40 | 1 | | T |
| 7 | 400 | 1 | | E |
| 8 | 399 | 1 | | E |
| 9 | 729 | 1 | | E |
| 10 | 913 | 1 | | E |
| 11 | 788 | 1 | | E |
| 12 | 145 | 1 | | E |
| 13 | 4 | 1 | | E |
| 14 | 797 | 1 | | E |
| 15 | 322 | 1 | | E |
| 16 | 902 | 1 | | E |
| 17 | 219 | 1 | | G |
| 18 | 64 | 1 | | G |
| 19 | 12 | 1 | | G |
| 20 | 8 | 1 | | G |

**Threshold scores**
Working towards the expected standard (T): 11 or fewer
Working at the expected standard (E): 12–15
Working at greater depth (G): 16–20

**Balance of difficulty of questions in the paper**
6 marks at working towards (T)
10 marks at the expected standard (E)
4 marks at working at greater depth (G)

# Year 3 Autumn Half Term 2: Reasoning Mark Scheme

| Question | Requirement | Mark | Additional guidance | Level of demand |
|---|---|---|---|---|
| 1 | Sixty-one | 1 | | T |
| 2 | 56 | 1 | | T |
| 3 | 30 | 1 | | T |
| 4 | True<br>False<br>True<br>True | 1 | | T |
| 5 | <table><tr><td>×</td><td>3</td><td>4</td><td>8</td></tr><tr><td>4</td><td>12</td><td>**16**</td><td>**32**</td></tr><tr><td>**5**</td><td>15</td><td>20</td><td>40</td></tr><tr><td>9</td><td>27</td><td>**36**</td><td>72</td></tr></table> | 2 | | E |
| 6 | 215 words | 2 | Award **2 marks** for the correct answer of 215 words.<br>If the answer is incorrect, award **1 mark** for evidence of an appropriate method, e.g.<br>950 − 330 − 405 OR<br>950 − (330 + 405). | E |
| 7 | **300** 350 400 450 500 **550**<br>400 500 **600** 700 800 **900** | 2 | | E |
| 8 | 6 bunches | 1 | | E |
| 9 | 2 × 6 = 12 = 4 × 3 | 1 | Accept 1 × 12.<br>Accept multiplications written in any order. | E |
| 10 | 27 counters | 2 | Award **2 marks** for the correct answer of 27 counters.<br>If the answer is incorrect, award **1 mark** for evidence of an appropriate method, e.g.<br>3 + (8 × 3) **or**<br>(8 × 3) + 3. | G |
| 11 | No. Some multiples are 3-digit numbers (200) but some multiples of 100 have more than 3 digits (1,200). | 1 | | G |

**Threshold scores**
Working towards the expected standard (T): 8 or fewer
Working at the expected standard (E): 9–11
Working at greater depth (G): 12–15

**Balance of difficulty of questions in the paper**
4 marks at working towards (T)
8 marks at the expected standard (E)
3 marks at working at greater depth (G)

© HarperCollins*Publishers* Ltd 2019

# Year 3 Spring Half Term 1: Arithmetic Mark Scheme

| Question | Requirement | Mark | Additional guidance | Level of demand |
|---|---|---|---|---|
| 1 | 141 | 1 | | T |
| 2 | 23 | 1 | | T |
| 3 | 39 | 1 | | T |
| 4 | 60 | 1 | | T |
| 5 | 66 | 1 | | T |
| 6 | 154 | 1 | | E |
| 7 | 700 | 1 | | E |
| 8 | 56 | 1 | | E |
| 9 | 101 | 1 | | E |
| 10 | 27 | 1 | | E |
| 11 | 100 | 1 | | E |
| 12 | 8 | 1 | | E |
| 13 | 630 | 1 | | E |
| 14 | 28 | 1 | | E |
| 15 | 7 | 1 | | E |
| 16 | 72 | 1 | | E |
| 17 | 963 | 1 | | E |
| 18 | 69 | 1 | | E |
| 19 | 168 | 1 | | E |
| 20 | 15 | 1 | | E |
| 21 | 15 | 1 | | G |
| 22 | 909 | 1 | | G |
| 23 | 140 | 1 | | G |
| 24 | 29 | 1 | | G |
| 25 | 18 | 1 | | G |

**Threshold scores**
Working towards the expected standard (T): 14 or fewer
Working at the expected standard (E): 15–19
Working at greater depth (G): 20–25

**Balance of difficulty of questions in the paper**
5 marks at working towards (T)
15 marks at the expected standard (E)
5 marks at working at greater depth (G)

# Year 3 Spring Half Term 1: Reasoning Mark Scheme

| Question | Requirement | Mark | Additional guidance | Level of demand |
|---|---|---|---|---|
| 1 | Box 3 ticked | 1 | | T |
| 2 | 19 − 3 = 16 | 1 | Accept 19 − 16 = 3. | T |
| 3 | 2 × 7 circled<br>7 + 7 circled<br>14 ÷ 2 circled | 1 | Accept alternative unambiguous indications e.g. ticked. | T |
| 4a | 24 people | 1 | | T |
| 4b | 6 cupcakes | 1 | | T |
| 5 | 868  866  688  686  668 | 1 | | E |
| 6 | £1  10p  2p<br>20p  20p  10p  50p  10p  2p | 2 | | E |
| 7 | 292 ←10 less— 302 —10 more→ 312<br>479 ←100 less— 579 —100 more→ 679 | 2 | Award **2 marks** for three correct numbers.<br>Award **1 mark** for two correct numbers. | E |
| 8a | The Magic Pineapple | 1 | Accept any unambiguous spelling, including MP. | E |
| 8b | 16 children | 1 | | E |
| 8c | 12 children | 1 | | E |
| 9 | 236 pence | 2 | Award **2 marks** for the correct answer of 236 pence.<br>If the answer is incorrect, award **1 mark** for evidence of an appropriate method, e.g.<br>£5 − £2 and 15 pence − 49 pence<br>or<br>£5 − (£2 and 15 pence + 49 pence)<br>Accept £2.36. | E |
| 10 | Eve<br>Explanation to show that:<br>• 5 × 2 = 10 or twice the height of a tower with 5 blocks would be 10 blocks<br>**or**<br>• 5 × 10 ≠ 10 or tens times the height of a tower with 5 blocks would not be 10 blocks<br>**or**<br>• 10 × 10 = 100 or tens times the height of a tower with 10 blocks would be 100 blocks. | 1 | | E |
| 11 | 320 marbles | 2 | Award **2 marks** for the correct answer of 320 marbles.<br>If the answer is incorrect, award **1 mark** for evidence of an appropriate method, e.g.<br>25 × 8 = 200<br>40 × 3 = 120<br>200 + 120 = | G |

© HarperCollins*Publishers* Ltd 2019

# Year 3 Spring Half Term 1: Reasoning Mark Scheme

| 12 | 1 5 [3]<br>+ [5] [6] 4<br>⎯⎯⎯⎯⎯<br>7 1 [7]<br>₁ | 2 | Award **2 marks** for three correct numbers.<br>Award **1 mark** for two correct numbers. | G |

**Threshold scores**
Working towards the expected standard (T): 11 or fewer
Working at the expected standard (E): 12–15
Working at greater depth (G): 16–20

**Balance of difficulty of questions in the paper**
5 marks at working towards (T)
11 marks at the expected standard (E)
4 marks at working at greater depth (G)

# Year 3 Spring Half Term 2: Arithmetic Mark Scheme

| Question | Requirement | Mark | Additional guidance | Level of demand |
|---|---|---|---|---|
| 1 | 80 | 1 | | T |
| 2 | 7 | 1 | | T |
| 3 | 12 | 1 | | T |
| 4 | 91 | 1 | | T |
| 5 | 972 | 1 | | T |
| 6 | 48 | 1 | | E |
| 7 | 367 | 1 | | E |
| 8 | 401 | 1 | | E |
| 9 | 940 | 1 | | E |
| 10 | 326 | 1 | | E |
| 11 | 72 | 1 | | E |
| 12 | 801 | 1 | | E |
| 13 | $\frac{4}{10}$ | 1 | | E |
| 14 | 8 | 1 | | E |
| 15 | 15 | 1 | | E |
| 16 | 905 | 1 | | E |
| 17 | 152 | 1 | | E |
| 18 | $\frac{8}{10}$ | 1 | | E |
| 19 | 7 | 1 | | E |
| 20 | 10 | 1 | | E |
| 21 | 600 | 1 | | G |
| 22 | 14 | 1 | | G |
| 23 | 10 | 1 | | G |
| 24 | 20 | 1 | | G |
| 25 | 60 | 1 | | G |

**Threshold scores**
Working towards the expected standard (T): 14 or fewer
Working at the expected standard (E): 15–19
Working at greater depth (G): 20–25

**Balance of difficulty of questions in the paper**
5 marks at working towards (T)
15 marks at the expected standard (E)
5 marks at working at greater depth (G)

# Year 3 Spring Half Term 2: Reasoning Mark Scheme

| Question | Requirement | Mark | Additional guidance | Level of demand |
|---|---|---|---|---|
| 1 | 15 ÷ 5 = ✓ | 1 | | T |
| 2 | $\frac{4}{10}$ | 1 | Accept $\frac{2}{5}$ or any equivalent. | T |
| 3 | Nineteen — 19, Twenty-nine — 29, Ninety-two — 92, Nine — 9 | 1 | | T |
| 4 | 46, 29 (→ 75); 65, 28 (→ 93) | 2 | Award **1 mark** for each correct missing answer. | T |
| 5 | 11 cm | 1 | | E |
| 6 | 20 children | 2 | Award **2 marks** for the correct answer of 20 children. If the answer is incorrect, award **1 mark** for evidence of an appropriate method, e.g. $32 - (\frac{1}{4} \text{ of } 32) - (\frac{1}{8} \text{ of } 32)$ or $32 - [(\frac{1}{4} \text{ of } 32) + (\frac{1}{8} \text{ of } 32)]$ | E |
| 7 | Pyramid: 128 / 16, 8 / 2, 8, 1 | 2 | | E |
| 8a | 50 | 1 | | E |
| 8b | 35 | 1 | Accept 34–36. | E |
| 8c | 85 | 1 | Accept 84–86. | E |
| 9 | 1 m 5 cm   1 m 70 mm   109 cm   1 m 15 cm | 2 | Accept Ben, Sam, Tia, Raj. Accept any unambiguous spellings, including B, S, T, R. | E |
| 10 | No, and then explanation should show that:<br>• 5 × 3 ≠ 25<br>or<br>• 5 × 3 = 15<br>or<br>• 25 ÷ 3 ≠ 5 or 8 r 1<br>or<br>• 25 ÷ 5 = 5 | 1 | Do not accept vague or incomplete explanations. Do not accept 'No' circled without a correct explanation. Accept 'Yes' or 'No' circled with the correct explanation. | E |

# Year 3 Spring Half Term 2: Reasoning Mark Scheme

| 11 | 792 cm<br>or 7 metres 92 cm | | 2 | Award **2 marks** for the correct answer of<br>7 metres 92 cm or 792 cm.<br>If the answer is incorrect, award **1 mark** for evidence of an appropriate method, e.g.<br>1200 – 300 – 108<br>**or**<br>1200 – (300 + 108). | G |
|---|---|---|---|---|---|
| 12 | 15<br>2 | | 2 | | G |

**Threshold scores**
Working towards the expected standard (T): 11 or fewer
Working at the expected standard (E): 12–15
Working at greater depth (G): 16–20

**Balance of difficulty of questions in the paper**
5 marks at working towards (T)
11 marks at the expected standard (E)
4 marks at working at greater depth (G)

# Year 3 Summer Half Term 1: Arithmetic Mark Scheme

| Question | Requirement | Mark | Additional guidance | Level of demand |
|---|---|---|---|---|
| 1 | 122 | 1 | | T |
| 2 | 21 | 1 | | T |
| 3 | 700 | 1 | | E |
| 4 | 805 | 1 | | E |
| 5 | 205 | 1 | | E |
| 6 | 251 | 1 | | E |
| 7 | 28 | 1 | | E |
| 8 | 5 | 1 | | E |
| 9 | 400 | 1 | | E |
| 10 | 24 | 1 | | E |
| 11 | 812 | 1 | | E |
| 12 | 100 | 1 | | E |
| 13 | 246 | 1 | | E |
| 14 | 5 | 1 | | E |
| 15 | 144 | 1 | | E |
| 16 | 9 | 1 | | E |
| 17 | 585 | 1 | | E |
| 18 | 1 | 1 | $\frac{7}{7}$ should be accepted also | E |
| 19 | 23 | 1 | | E |
| 20 | $\frac{3}{6}$ | 1 | Accept also $\frac{1}{2}$ or 0.5 | E |
| 21 | 295 | 1 | | E |
| 22 | 222 | 1 | | E |
| 23 | 8 | 1 | | E |
| 24 | 12 | 1 | | E |
| 25 | 199 | 1 | | G |
| 26 | 2 | 1 | | G |
| 27 | $\frac{3}{9}$ | 1 | Accept $\frac{1}{3}$ | G |
| 28 | $\frac{6}{8}$ | 1 | Accept $\frac{3}{4}$ | G |
| 29 | 45 | 1 | | G |
| 30 | 28 | 1 | | G |

**Threshold scores**
Working towards the expected standard (T): 17 or fewer
Working at the expected standard (E): 18–23
Working at greater depth (G): 24–30

**Balance of difficulty of questions in the paper**
2 marks at working towards (T)
22 marks at the expected standard (E)
6 marks at working at greater depth (G)

© HarperCollins*Publishers* Ltd 2019

# Year 3 Summer Half Term 1: Reasoning Mark Scheme

| Question | Requirement | Mark | Additional guidance | Level of demand |
|---|---|---|---|---|
| 1 | (bar model shaded) | 1 | Accept alternative unambiguous positive indications e.g. circled. | T |
| 2 | 7 | 1 | | T |
| 3 | 10p 10p 20p 20p 10p 5p<br>10p 10p 10p 10p 10p 10p 10p 5p<br>20p 50p 2p 2p 1p | 1 | All three boxes must be ticked for the award of **1 mark**. | T |
| 4 | 67 \| 20 \| 13<br>30 \| 45 \| 25<br>3 \| 35 \| 62 | 2 | Award **2 marks** for four correct numbers.<br>Award **1 mark** for two or three correct numbers. | T |
| 5 | $1 \quad \frac{7}{8} \quad \frac{4}{8} \quad \frac{2}{8}$ | 1 | | E |
| 6 | $\frac{2}{10}$<br>$\frac{8}{10} = \frac{4}{5}$ | 2 | | E |
| 7 | There are 24 hours in a day.<br>There are 12 months in a year.<br>There are 60 minutes in a hour.<br>There are 365 days in a year.<br>There are 366 days in a leap year. | 2 | | E |
| 8a | twenty past ten | 1 | Accept 20 past 10 **or** ten twenty. | E |
| 8b | 7:42 (p.m.) | 1 | | E |
| 8c | nine fifteen (p.m.)<br>**or**<br>quarter past nine (in the evening) | 1 | Accept $\frac{1}{4}$ past 9 **or** 15 past 9 **or** nine fifteen | E |
| 9 | 231 T-shirts | 2 | Award **2 marks** for the correct answer of 231 T-shirts.<br>If the answer is incorrect, award **1 mark** for evidence of an appropriate method, e.g.<br>151 − 147 = 4<br>235 − 4 = | E |
| 10a | 3:52 (p.m.) | 1 | Accept 03:52 or 15:52 | E |
| 10b | 5 seconds | 1 | | E |
| 11 | $\frac{5}{12}$ | 2 | Award **2 marks** for the correct answer of 5 slices.<br>If the answer is incorrect, award **1 mark** for evidence of an appropriate method, e.g.<br>$1 - (\frac{4}{12} + \frac{3}{12}) =$ | E |
| 12 | three hundred and five | 1 | | E |
| 13 | 2 flower pots | 2 | Award **2 marks** for the correct answer of 2 flower pots.<br>If the answer is incorrect, award **1 mark** for evidence of an appropriate method, e.g.<br>(78 ÷ 3) − (96 ÷ 4) = | G |

© HarperCollins*Publishers* Ltd 2019

# Year 3 Summer Half Term 1: Reasoning Mark Scheme

| 14a | $\frac{7}{9} + \frac{2}{9} = 1$ or $\frac{2}{9} + \frac{7}{9} = 1$ | 1 | | G |
|---|---|---|---|---|
| 14b | $1 - \frac{2}{9} = \frac{7}{9}$ or $1 - \frac{7}{9} = \frac{2}{9}$ | 1 | | G |
| 15 | An explanation that shows A choice of $\frac{1}{3}$ of (£)45 supported by a reason, e.g. $\frac{1}{3}$ of (£)45 = (£)15 $\frac{3}{8}$ of (£)24 = (£)9 and (£)15 > (£)9 | 1 | Accept an acceptable decision to choose the smaller amount if supported by correct calculation. | G |

**Threshold scores**
Working towards the expected standard (T): 14 or fewer
Working at the expected standard (E): 15–19
Working at greater depth (G): 20–25

**Balance of difficulty of questions in the paper**
5 marks at working towards (T)
15 marks at the expected standard (E)
5 marks at working at greater depth (G)

# Year 3 Summer Half Term 2: Arithmetic Mark Scheme

| Question | Requirement | Mark | Additional guidance | Level of demand |
|---|---|---|---|---|
| 1 | 385 | 1 | | E |
| 2 | 899 | 1 | | E |
| 3 | 181 | 1 | | E |
| 4 | 16 | 1 | | E |
| 5 | 250 | 1 | | E |
| 6 | 662 | 1 | | E |
| 7 | 480 | 1 | | E |
| 8 | 88 | 1 | | E |
| 9 | 6 | 1 | | E |
| 10 | 8 | 1 | | E |
| 11 | 883 | 1 | | E |
| 12 | $\frac{7}{8}$ | 1 | | E |
| 13 | 189 | 1 | | E |
| 14 | 421 | 1 | | E |
| 15 | 432 | 1 | | E |
| 16 | 8 | 1 | | E |
| 17 | 25 | 1 | | E |
| 18 | 4 | 1 | | E |
| 19 | $\frac{9}{10}$ | 1 | | E |
| 20 | 200 | 1 | | E |
| 21 | 1173 | 1 | | E |
| 22 | 205 | 1 | | E |
| 23 | 36 | 1 | | E |
| 24 | 14 | 1 | | E |
| 25 | 6 | 1 | | G |
| 26 | 1099 | 1 | | G |
| 27 | $\frac{4}{6}$ | 1 | Accept $\frac{2}{3}$ | G |
| 28 | 1 | 1 | | G |
| 29 | 40 | 1 | | G |
| 30 | $\frac{6}{10}$ | 1 | Accept $\frac{3}{5}$ and 0.6 | G |

© HarperCollins*Publishers* Ltd 2019

# Year 3 Summer Half Term 2: Arithmetic Mark Scheme

**Threshold scores**
Working towards the expected standard (T): 17 or fewer
Working at the expected standard (E): 18–23
Working at greater depth (G): 24–30

**Balance of difficulty of questions in the paper**
0 marks at working towards (T)
24 marks at the expected standard (E)
6 marks at working at greater depth (G)

# Year 3 Summer Half Term 2: Reasoning Mark Scheme

| Question | Requirement | Mark | Additional guidance | Level of demand |
|---|---|---|---|---|
| 1 | <table><tr><td>+</td><td>16</td><td>**45**</td><td>68</td></tr><tr><td>25</td><td>**41**</td><td>70</td><td>**93**</td></tr><tr><td>**27**</td><td>43</td><td>72</td><td>95</td></tr></table> | 2 | Award **2 marks** for all 4 correct missing numbers. Award **1 mark** for 3 correct missing numbers. | T |
| 2 | 6 packs | 2 | Award **2 marks** for the correct answer of 6 packs. If the answer is incorrect, award **1 mark** for evidence of an appropriate method, e.g. (30 ÷ 10) + (15 ÷ 5) **or** (15 ÷ 5) + (30 ÷ 10) | T |
| 3 | Scales → kg; jug → ml; thermometer → °C; tape measure → cm | 1 | | T |
| 4 | Arrow drawn on scale showing approximately 350 g | 1 | Allow for an inaccurate drawing, accept an arrow showing a mass > 325 g and < 375 g. | E |
| 5 | $\frac{3}{10}$<br>$\frac{7}{10}$<br>$1\frac{1}{10}$ | 1 | | E |
| 6 | Yes, Jack, is correct followed by an explanation that shows any property of a cuboid, e.g.<br>• It has 6 rectangular faces<br>• It has 12 edges that meet at right angles. | 1 | Accept also two square faces and four rectangular faces. | E |
| 7a | 150 ml | 1 | | E |
| 7b | 700 ml | 1 | | E |
| 8 | 316 suitcases | 2 | Award **2 marks** for the correct answer of 316. If the answer is incorrect, award **1 mark** for evidence of an appropriate method, e.g. 678 − 234 − 128 = **or** 678 − (234 + 128) = | E |
| 9 | <table><tr><td></td><td>At least one right angle</td><td>No right angles</td></tr><tr><td>4-sided</td><td>rectangle</td><td>trapezium</td></tr><tr><td>not 4-sided</td><td>right triangle</td><td>circle</td></tr></table> | 2 | Accept any shapes that meet the given criteria:<br>• a quadrilateral with one right angle<br>• a shape that does not have 4 sides and does not have a right angle. | E |

# Year 3 Summer Half Term 2: Reasoning Mark Scheme

| | | | | |
|---|---|---|---|---|
| 10a | $2\frac{1}{2}$ kg; 2075 g; 1 kg 250 g | 1 | | E |
| 10b | 1,250 grams | 1 | Accept 1.25 kg or 1 kg 250 g. | E |
| 11 | F | 1 | | E |
| 12a | Hundreds: 100, 100, 100, 100; Tens: (empty); Ones: 1, 1, 1, 1, 1, 1, 1, 1, 1 | 1 | Accept 409 drawn correctly, using place value counters or dienes. Do not accept 4 and 9 written on the chart. | E |
| 12b | < ; < | 1 | | E |
| 13 | Both Jared and Hannah are correct. A quarter turn anticlockwise and a three-quarter turn clockwise would reach the same position. | 1 | Accept a diagram showing that the turns have the same end point. | E |
| 14 | 32 g | 2 | Award **2 marks** for the correct answer of 32 g. If the answer is incorrect, award **1 mark** for evidence of an appropriate method, e.g. (196 − 100) ÷ 3 = | G |
| 15 | 5 3 × 2 | 1 | | G |
| 16a | 35 girls | 1 | Accept an answer in the range 34–36. | G |
| 16b | 15 boys | 1 | Accept an answer in the range 14–16. | G |

**Threshold scores**
Working towards the expected standard (T): 14 or fewer
Working at the expected standard (E): 15–19
Working at greater depth (G): 20–25

**Balance of difficulty of questions in the paper**
5 marks at working towards (T)
15 marks at the expected standard (E)
5 marks at working at greater depth (G)

# Content domain references

| Autumn 1: Arithmetic ||
|---|---|
| Question | Content domain reference |
| 1 | 2C2a |
| 2 | 2C1 |
| 3 | 2C2a |
| 4 | 2C2a |
| 5 | 2C3 |
| 6 | 2C3 |
| 7 | 2C2b |
| 8 | 2C6 |
| 9 | 3N2b |
| 10 | 3N2b |
| 11 | 3C2 |
| 12 | 3C2 |
| 13 | 3C2 |
| 14 | 3C2 |
| 15 | 3C2 |
| 16 | 3C2 |
| 17 | 3C2 |
| 18 | 3C1 |
| 19 | 3C2 |
| 20 | 3C4 |

| Autumn 2: Arithmetic ||
|---|---|
| Question | Content domain reference |
| 1 | 2C2a |
| 2 | 2C2a |
| 3 | 2C2a |
| 4 | 2C6 |
| 5 | 2C6 |
| 6 | 2C2a |
| 7 | 3C1 |
| 8 | 3N2b |
| 9 | 3C2 |
| 10 | 3C2 |
| 11 | 3C2 |
| 12 | 3C2 |
| 13 | 3C7 |
| 14 | 3C2 |
| 15 | 3C2 |
| 16 | 3C2 |
| 17 | 3C2 |
| 18 | 3C7 |
| 19 | 3C7 |
| 20 | 3C7 |

| Autumn 1: Reasoning ||
|---|---|
| Question | Content domain reference |
| 1 | 2N2b |
| 2 | 2C2b |
| 3 | 2N6 |
| 4 | 2N1 |
| 5 | 3C3 |
| 6 | 3N2a/3N2b |
| 7 | 3C2 / 3C4 |
| 8a | 3N4 / 3C2 / 3C4 |
| 8b | 3N4 / 3C2 / 3C4 |
| 9 | 3N2a / 3N6 |
| 10 | 3C2 / 3C4 |
| 11 | 3N3 / 3N4 |

| Autumn 2: Reasoning ||
|---|---|
| Question | Content domain reference |
| 1 | 2N3 / 2N4 |
| 2 | 2C2b / 2C4 |
| 3 | 2C1/2C3 |
| 4 | 2N2b / 2N2a / 2N4 |
| 5 | 3C6/3C8 |
| 6 | 3C2/3C4 |
| 7 | 3N1b |
| 8 | 3C8 |
| 9 | 3C7/3C8 |
| 10 | 3C8 |
| 11 | 3N1b |

# Content domain references

| Spring 1: Arithmetic | |
|---|---|
| Question | Content domain reference |
| 1 | 2C2a |
| 2 | 2C2a |
| 3 | 2C3 |
| 4 | 2C6 |
| 5 | 2C3 |
| 6 | 3C1 |
| 7 | 3C2 |
| 8 | 3C6 |
| 9 | 3C2 |
| 10 | 3C6 |
| 11 | 3N2b |
| 12 | 3C6 |
| 13 | 3N2b |
| 14 | 3C2 |
| 15 | 3C6 |
| 16 | 3C6 |
| 17 | 3C2 |
| 18 | 3C7 |
| 19 | 3C7 |
| 20 | 3C7 |
| 22 | 3C7 |
| 23 | 3N2b |
| 23 | 3C7 |
| 24 | 3C7 |
| 25 | 3C8 |

| Spring 2: Arithmetic | |
|---|---|
| Question | Content domain reference |
| 1 | 2C2b |
| 2 | 2C6 |
| 3 | 2C6 |
| 4 | 3N2b |
| 5 | 3C1 |
| 6 | 3C6 |
| 7 | 3C2 |
| 8 | 3C2 |
| 9 | 3C2 |
| 10 | 3C2 |
| 11 | 3C6 |
| 12 | 3N2b |
| 13 | 3F1a |
| 14 | 3C8 |
| 15 | 3C7 |
| 16 | 3N2b |
| 17 | 3C7 |
| 18 | 3F1a |
| 19 | 3F1c |
| 20 | 3F1a |
| 21 | 3C1 |
| 22 | 3F1c |
| 23 | 3F1a |
| 24 | 3F1c |
| 25 | 3F1c |

| Spring 1: Reasoning | |
|---|---|
| Question | Content domain reference |
| 1 | 2S1/2S2b |
| 2 | 2C3 |
| 3 | 2C7 |
| 4a | 2C2b / 2C4 |
| 4b | 2C2b / 2C4 |
| 5 | 3N2a |
| 6 | 3M9a |
| 7 | 3N2b |
| 8 | 3S1 |
| 9 | 3M9a |
| 10 | 3C8 |
| 11 | 3C7/3C8 |
| 12 | 3C2 |

| Spring 2: Reasoning | |
|---|---|
| Question | Content domain reference |
| 1 | 2C8 |
| 2 | 2F1a |
| 3 | 2N2a |
| 4 | 2C2b/2C4 |
| 5 | 3M2a |
| 6 | 3F1c/3F1b / 3F10 |
| 7 | 3C6 |
| 8 | 3S1 |
| 9 | 3M1a / 3S1 |
| 10 | 3M7 |
| 11 | 3M9b |
| 12 | 3F1c/3F1b / 3F10 |

# Content domain references

| Summer 1: Arithmetic | |
|---|---|
| Question | Content domain reference |
| 1 | 2N2b |
| 2 | 2C2a |
| 3 | 3C1 |
| 4 | 3N2b |
| 5 | 3N2b |
| 6 | 3C2 |
| 7 | 3C6 |
| 8 | 3C6 |
| 9 | 3C2 |
| 10 | 3C6 |
| 11 | 3C2 |
| 12 | 3N2b |
| 13 | 3C7 |
| 14 | 3F1a |
| 15 | 3C7 |
| 16 | 3F1c |
| 17 | 3C2 |
| 18 | 3F4 |
| 19 | 3C7 |
| 20 | 3F4 |
| 21 | 3N2b |
| 22 | 3C4 |
| 23 | 3F1c |
| 24 | 3C8 |
| 25 | 3N2b |
| 26 | 3F1a |
| 27 | 3F4 |
| 28 | 3F4 |
| 29 | 3F1c |
| 30 | 3C8 |

| Summer 1: Reasoning | |
|---|---|
| Question | Content domain reference |
| 1 | 2F1a / 2F1c |
| 2 | 2C8 |
| 3 | 2M3a / 2M3b |
| 4 | 2C1/2C4 |
| 5 | 3F3 |
| 6 | 3F2/3F10 |
| 7 | 3M4e |
| 8 | 3M4a / 3M4d |
| 9 | 3C2/ 3C4 |
| 10a | 3M4f |
| 10b | 3M4d |
| 11 | 3F4/ 3F10 |
| 12 | 3N2a/3N4 |
| 13 | 3C7 |
| 14 | 3F4/ 3F10 |
| 15 | 3F1c/3F1b / 3F10 |

| Summer 2: Arithmetic | |
|---|---|
| Question | Content domain reference |
| 1 | 3N2b |
| 2 | 3N2b |
| 3 | 3C2 |
| 4 | 3C6 |
| 5 | 3C2 |
| 6 | 3C2 |
| 7 | 3C2 |
| 8 | 3C6 |
| 9 | 3F1a |
| 10 | 3C6 |
| 11 | 3C2 |
| 12 | 3F4 |
| 13 | 3C7 |
| 14 | 3C2 |
| 15 | 3C7 |
| 16 | 3F1c |
| 17 | 3C7 |
| 18 | 3C8 |
| 19 | 3F1a |
| 20 | 3C1 |
| 21 | 3C2 |
| 22 | 3C2 |
| 23 | 3F1c |
| 24 | 3C7 |
| 25 | 3C8 |
| 26 | 3N2b |
| 27 | 3F4 |
| 28 | 3F1a |
| 29 | 3F1c |
| 30 | 3F4/3F10 |

| Summer 2: Reasoning | |
|---|---|
| Question | Content domain reference |
| 1 | 2C1/2C4 |
| 2 | 2C8 |
| 3 | 2M2 |
| 4 | 3M2b |
| 5 | 3F1a |
| 6 | |
| 7a | 3M1c/3M2c |
| 7b | 3M9d |
| 8 | 3C2 / 3C4 |
| 9 | 3G4b |
| 10a | 3M1b |
| 10b | 3M9c |
| 11 | 3G2c |
| 12 | 3N4 / 3N2a |
| 13 | 3G3b |
| 14 | 3M9c |
| 15 | 3C7 |
| 16 | 3S1 |

Name _____     Class _____

## Year 3/P4 Maths Progress Tests for White Rose Record Sheet

| Tests | Mark | Total marks | Key skills to target |
|---|---|---|---|
| Autumn 1: Arithmetic | | | |
| Autumn 1: Reasoning | | | |
| Autumn 2: Arithmetic | | | |
| Autumn 2: Reasoning | | | |
| Spring 1: Arithmetic | | | |
| Spring 1: Reasoning | | | |
| Spring 2: Arithmetic | | | |
| Spring 2: Reasoning | | | |
| Summer 1: Arithmetic | | | |
| Summer 1: Reasoning | | | |
| Summer 2: Arithmetic | | | |
| Summer 2: Reasoning | | | |

© HarperCollins*Publishers* Ltd 2019